COGNITION AND SAFETY

T0326181

Cognition and Safety
An Integrated Approach to
Systems Design and Assessment

OLIVER STRÄTER

Safety and Security Management, EUROCONTROL, Belgium
and
Institute of Ergonomics, Institute of Technology Munich, Germany

Routledge
Taylor & Francis Group

LONDON AND NEW YORK

First published 2005 by Ashgate Publishing

2 Park Square, Milton Park, Abingdon, Oxon OX14 4RN
711 Third Avenue, New York, NY 10017, USA

Routledge is an imprint of the Taylor & Francis Group, an informa business

First issued in paperback 2016

British Library Cataloguing in Publication Data
Sträter, Oliver
   Cognition and safety : an integrated approach to systems
   design and assessment
   1.Industrial safety - Psychological aspects 2.Cognitive
   psychology
   I.Title
   363.1'172

Library of Congress Control Number: 2005921529

ISBN 978-0-7546-4325-8 (hbk)
ISBN 978-1-138-26671-1 (pbk)

Transfered to Digital Printing in 2009

Contents

List of Figures

List of Tables

Abbreviations

ACT: Adaptive Control of Thought
AEOD: Analysis and Evaluation of Operation Data
AM: Accident Management
ARAS: Ascending Reticular Activation System
ATHEANA: A Technique for Human Error Analysis
ATM: Air Traffic Management

CAHR: Connectionism Assessment of Human Reliability
CCT: Cognitive Complexity Theory
CES: Cognitive Environmental Simulation
CM: Confusion Matrix approach
COCOM: Contextual Control Model
COSI: Cognitive Simulation
COSIMO: Cognitive Simulation Model
CPU: Central Processing Unit
CREAM: Cognitive Reliability and Error Analysis Method

EEG: Electroencephalogram
EFC: error-forcing context
EOC: Error of Commission
ESAT: Expertensystem zur Aufgaben-Taxonomie (Expert-System for Task Taxonomy)

FRAM: Functional Resonance Accident Models

GEMS: Generic Error Modelling System
GPWS: Ground Proximity Warning System
GRS: Gesellschaft für Anlagen- und Reaktorsicherheit

HCR: Human Cognitive Reliability Model
HEART: Human Error Assessment and Reduction Technique
HEP: Human Error Probability
HERA: Human Error in Air traffic management

IDAC: Information, Decision, Action in Crew context
INTENT: Method for estimating human error probabilities for errors of Intention

LTM: Long-Term Memory

MERMOS: Méthode d'Evaluation de la Réalisations des Missions Opérateur pour la Sûreté
MIDAS: Man Machine Integration Design and Analysis System
MMS: Man Machine System
MRT: Multiple Resource Theory

NARA: Nuclear Action Reliability Assessment

OPPS/SAINT: System Analysis Integrated Networks of Tasks

SLIM: Success Likelihood Index Method
SOAR: Unified models of cognition
STM: Short-Term Memory

TEMM: Threat-Error-Management Model
THERP: Technique for Human Error Rate Prediction
VR: Virtual Reality

Preface

Severe accidents, incidents and daily safety practice in any industry show the importance of human behaviour for safety. Throughout many industries, quite similar observations can be made regarding the problems of integrating human performance characteristics into design, investigating incidents and assessing human performance.

Design and operation of systems use different methods and models to deal with this important aspect of system safety. For instance, operation is concerned with human error, while design aligns the system to workload or situational awareness. Owing to this methodological gap, a homogeneous treatment of human performance in system design and operation becomes difficult.

The main idea of this book is to focus on an integrated view of cognitive human issues, which means that concepts such as human error, workload, or situational awareness are described in a homogeneous approach. It exploits and integrates findings from physiological and psychological research and applied studies. This provides the reader with a generic perspective of cognitive issues, which s/he then can easily adjust to their own problems. It also provides the reader with a homogeneous view on the variety of cognitive approaches, which usually do not fit well together or have unclear interfaces to each other. Such variety makes it difficult for the practitioner to consider the broad spectrum of cognitive aspects and to improve safety herewith. Another aspect tackled in this book is that cognitive aspects will only be considered if experienced in daily work on future system design. The book will therefore suggest an integrated perspective of cognition in technical systems, which is needed to implement this important field into the design of systems and organizations.

An integrated model provides the opportunity to combine data collection efforts of different industries regarding human performance and to enhance the cross-boundary learning and improvement of systems regarding human characteristics. Examples from process industry, nuclear, aviation, air traffic management, automobile, manufacturing, and medicine are provided.

'Good data means no worry' (Carnegie, 1999)

Acknowledgements

The book covers back to several years of work in various safety-related industries and projects, in aviation and air traffic management, nuclear, automobile and occupational safety. The author would like to pass special thanks to those who helped generate the thoughts reflected in this book. Special thanks are given to:

- the System Ergonomics Group of the Institute of Ergonomics of the University of Technology Munich;
- the Siemens occupational safety department with the Ergonomics Competence Network (E-C-N);
- the project on Group Interaction in High Risk Environments (GIHRE);
- the OECD Working Group Risk;
- the OECD Halden Reactor Project;
- the members of the German Working Group on Human Reliability of the VDI (Verein Deutscher Ingenieure; Association of German Engineers);
- the members of Eurocontrol's project on Solutions for Human Automation Partnership in European Air Traffic Management (SHAPE);
- the joint working group on errors of commission of Electricité de France, Paul Scherrer Institute/Switzerland, KEMA/Netherlands, and GRS/Germany;
- the ‚Gesellschaft für Anlagen- und Reaktorsicherheit' (GRS);
- the German Reactor Safety Commission (RSK);
- the Eurocontrol Safety and Security Management Unit;
- the Eurocontrol Human Factors and Manpower Unit;
- the Eurocontrol Experimental Centre.

Part I

Concerns

Chapter 1

The Need to Model Cognition in Safety

The Challenge of Cognition for Safe Design and Organization of Systems

Human society has become an information processing society. Virtually any industrial area is currently challenged by the impact of new technologies on human cognition. A car becomes much more of a computer by the increased use of navigation. Surveillance systems and power plants can be operated by a single person, and aircraft and airspace can be managed much more efficiently by the increased use of computer technology. New technologies also change the organizational structures of the industries. As the Internet provides communication around the world, staff become more remote and distributed. Reduced permanent staff in the operation of industrial systems goes hand-in-hand with increased hiring of external staff for maintenance or system development. Team-working and communication suffer from such developments (Sträter et al., 2003).

Disasters show that something is wrong in a system, and the list of disasters is long. Examples include

- the Challenger explosion in 1986,
- the fire on the Piper Alpha oil-platform in 1988,
- the flight into terrain of an Airbus 320 in 1992,
- the reactor explosion in Chernobyl in 1986,
- the core melt of the TMI nuclear power plant in 1979,
- the release of radioactivity in Tokai Mura in 1997, or
- the mid-air collision at Lake Constance in 2002.

These accidents will be described in further detail in the course of this book, particularly regarding the cognitive aspects involved.

Besides such events with a high public perception, daily events show system deficiencies as well. The costs for occupational accidents in Germany for the year 2002 were about € 44.15 billion (BAUA, 2002). According to Helmreich (GIHRE, 2004) about 44,000 to 98,000 casualties were encountered in 2002 in US-hospitals due to inappropriate human actions in medical treatment and not due to illness itself. The study also refers to 2.7 errors per patient per day. The probability of receiving serious damage was observed as approximately 3.7% per patient. Night shifts were revealed as one of the most serious problems.

All accidents show weaknesses in design, maintenance or management, which affects the cognitive decision-making in operation, the sharp end of a system.

Typically, those weaknesses are latently present for a long time before an accident happens (Reason, 1997). Often they are known and tolerated because one states:

(a) Nothing happened so far in my system => My system is safe.

However, the conclusion that nothing happens based on the experience so far is an invalid counter-conclusion from the rule:

(b) My system is safe => Nothing will happen in my system.

Statement b) is the basic rule of inference (modus pollens) of safety engineering. However, disasters show that statement b) may never be achieved. Any accident shows that the rule of negative inference (modus tollens) needs to be applied to both statements. The rule of negative inference states that the only valid solution of statement a) or b) for any system is (Popper, 1997):

(c) There are unsafe elements in my system.

Operational Levels

Independently from the industry, the humans at the front end of an operation are often identified as the cause of the malfunction and as the unsafe element, because they intervene directly with the technical system and are performing the actual action in incidents and accidents that breaks down the system. The term *Human Factor* was coined to describe the human role in operation. However, detailed analyses of the events usually show that other parties play a significant role in the cause and the development of the accidents. The staff at the operating level are requested to deal with many constraints stemming from several areas they have no influence on. The following operational levels may be distinguished as important contributors to safety (cf. Eurocontrol-Agenda, 2005):

- The working-level of the staff that directly deals with the technical system. Staff at this level usually 'perform' the error (so-called active error; Reason, 1990). However, the operational staff are not necessarily responsible or guilty for the errors performed.
- The maintenance-level. Level of the staff that maintain the system: Technical systems never work without maintenance. Consequently, maintenance actions may induce errors into the system, which are not apparent and do not directly lead to erroneous system states (latent errors according to Reason, 1990).
- The organizational-level of the staff that organize the tasks at the working level. Every system needs staff and resource planning besides the working level. Staff at this level are also exposed to potential erroneous behaviour. Typically, such errors may be due to decisions between safety and effectiveness.

- The design-level. Level of the staff that design the technical system. Errors in design are critical because they are latent and will probably not be mitigated because of the involved costs in doing so (Sträter & Bubb, 2003).
- The regulator-level of the staff that regulate the system. In the highly complex world of today's systems, extensive regulations are used to ensure that the systems and the staff meet the requirements for safety operation (e.g. driver licenses are a simple example). Certainly, errors can also be made at this level, which may cause unnecessary burdens at the working level and which finally leads indirectly to active errors. Overruling is a well-known problem.

These levels depend on each other in a complex way. According to Leveson (2002), Figure 1.1 depicts the human influences on system safety, covering the whole range from system development to system operations at all involved working levels.

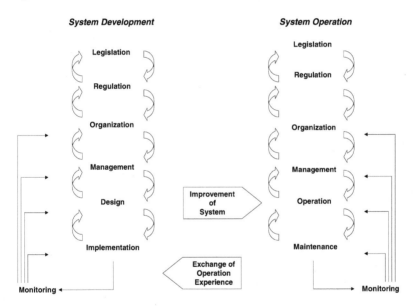

Figure 1.1   Interrelation of operational levels from design-level to working-level (adapted from Leveson, 2002)

Furthermore, the discussion of the different levels indicates that the human influence cannot be eliminated or reduced by automation. The attempt to reduce the human impact by introducing automation merely shifts the human influence to other than the working level. Bainbridge (1987) entitled the fallacy that automation reduces the relevance of humans for system safety as *Ironies of Automation*. Automation simply changes the spectrum of persons involved in safety (from operators to technicians, managers, designers and regulators).

Considering increasing complexity and by longer-term impacts, one needs to consider safety impacts of the scientific and societal level. By defining paradigms and social recognition of a working environment, they play an important role in understanding human behaviour and error. For instance, the limited computer power in the early days of risk assessment did not allow for dynamic risk modelling approaches. The scientific community therefore decided to let fault-tree and event-tree approaches, which have inherent limitations, represent the dynamic aspects of human behaviour (see Part III of this book). Society in Germany, for instance, does currently not accept nuclear power. As a consequence, the regulation and operation suffer from the constraint that, any problem, even a slight one, in a nuclear plant is pilloried, which puts the operators under an enormous additional threat.

Cognitive Aspects of System Safety

The link of cognitive aspects to the working level   Automated systems have a specified range of operation (design-bases or design specification). However, modus tollens also applies to automation. There is always the possibility that systems operate beyond their specification (beyond design-bases), either by internal failures, interference / interdependence with other automated systems, or by operating in situations they were not designed for. The user at the working level has to judge at any time, whether the automation is working properly or beyond specification. Automation therefore enhances the importance of decision-making and diagnosis in system safety as judgements and decisions are required at the working level on the status of the automated systems. Procedures for system failures are designed to support the user in beyond design-bases situations. With a growing number of automated systems, the number of procedures also increases. As a downside of procedural support in cases of failures, decisions need to be made on the status of the automated systems as well as on the applicability of procedures. Automation therefore increases the variety of contextual conditions under which decisions are to be made. These constraints need to be resolved, which requires prioritization, utilization and goal setting in cognitive processing (Montmollin, 1992; Mosneron-Dupin et al., 1997).

Humans at the working level are forced to make decisions based on constraints from targets set at the management level, the procedures and interfaces given, the required communication with working partners and the operational tasks to be performed. This leads to the phenomena of induced mental workload. The term 'induced' comprises the additional effort due to the type of interaction with the system. A frequently stated selling argument of automation is that it reduces workload. However, induced workload may cause an even higher net workload for the user than the workload an automated system is designed to reduce.

Cognitive psychology consequently becomes a considerable contribution to ensure safety at the working level. Examples of the importance are frequent and only two are mentioned here. In the field of Air Traffic Management (ATM), the new technology and the organizational changes following from the new capabilities (such as airspace harmonization in Europe) lead to an increased development of tools and methods in order to compensate for the impact of new technology on the controller (Sträter et al., 2004a). Distractions from car driving by advanced navigation systems have become a critical topic for the realization of such systems in transport (Schweigert, 2003).

Overall, the mental demand of the user at the working level is increasing (Figure 1.1). Increased mental demand changes the type of errors. Errors of commission or violation by the staff at the working level are an unavoidable downside of automation (or of complex systems in general). Automation, if not designed properly, does not eliminate human errors at the working level, but changes them (Bainbridge, 1987). However, the causes for erroneous human behaviour can be found in the levels preceding the working level.

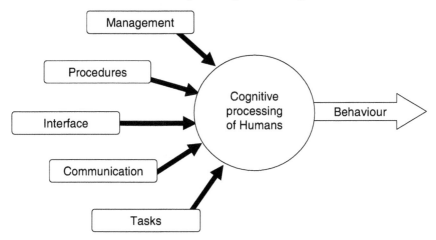

Figure 1.2   Constraints on decision-making and induced mental workload

The link of cognitive aspects to the organizational level According to the discussion in the preceding section, cognitive aspects of system safety stem from constraints, under which humans operate at the working level. As these constraints may not only be of a negative nature, the term *Context* was introduced for those conditions triggering certain human information processing behaviours (Hollnagel, 1998). The consequences from context regarding the cognitive demands are well known. Highly automated industries such as Aviation and Nuclear power plants have suffered from the problem for a long time while others such as ATM or the automobile industry are currently experiencing the effects.

Generally, one can observe the portion of human induced accidents increasing. Often it is stated that the proportion is getting higher, because the technical equipment itself is becoming increasingly reliable and therefore necessarily increases the proportion of human impacts. However, the discussion of induced workload above shows that this cannot be the only reason. Technology may induce certain types of human error into the system, if it is not designed properly. In particular, those human errors occurring due to problems of human-machine interaction are often incorrectly assigned as errors of the humans at the working level.

Triggering cognitive mechanisms, the context plays a major role in understanding and preventing events. The reasons for inappropriate contextual conditions can often be identified at the organizational level. During the Challenger disaster, for instance, NASA was under public pressure to have a successful mission. The management therefore decided to launch the shuttle, although serious safety problems regarding the sealing of the hydrogen tanks were known internally. Another space shuttle accident, the Columbia explosion, where the shuttle lost a tile during the start and exploded due to overheating in the landing-phase, was investigated by Woods (2003). He summarizes the classic patterns of cognitive aspects on the organizational level involved as follows:

- Drift toward failure as defences erode in the face of production pressure.
- An organization that takes past success as a reason for confidence instead of investing in anticipating the changing potential for failure.
- Fragmented problem solving process that clouds the big picture.
- Failure to revise assessments as new evidence accumulates.
- Breakdowns at the boundaries of organizational units that impede communication and coordination.

These patterns are widely independent from the industrial setting and the organizational level. As Woods manifests further, an organization is always a complex interdependent relationship between individuals with different backgrounds and experiences (a multi-disciplinary group of persons). The cognitive mechanism of these individuals is essential for understanding the organizational malfunctions. Organizationally induced accidents or events have their origins in cognitive issues (like misjudgements or misinterpretations) of the staff working at the organizational level. Therefore, cognitive decisions are strongly coupled to organizational influences on safety. Safety culture, safety management and knowledge management are the means currently discussed to improve the situation in these fields.

However, all these means will fail to be effective if the cognitive aspects of the persons involved in the inappropriate organization are not faced. Both the Challenger and the Columbia accident show that Figure 1.2 (above) also holds in principle for the organizational and regulatory levels of system operations. For example, management may be forced to make decisions based on their procedures (policies), interfaces, economic constraints and legal impairments. As well as the Challenger accident, the mid-air collision at Lake Constance and the nuclear accident in Tokai Mura are examples for the relevance of this influence.

The link of cognition important to the design level. Accidents are only the spectacular side of ill-operated, ill-designed or ill-managed systems. Problems occur a long time before an accident happens. Design influences the behaviour at the working level. Bad alignment of activities from design to operation has led to a considerable delay in system implementation of related costs in ATM. It also contributed to unwanted events. Regarding the space-shuttle disasters of Challenger in 1986 and Columbia in 2003 or the Alitalia crash in Zurich 1991, the required information about design problems with safety impacts had not been adequately processed over the years. Important information was missing at the working level and the system was not adjusted to safety needs.

In addition to the organizational issues discussed above, design issues disclose an additional aspect of the nature of accidents. Different parties are involved in the system design and these parties have different attitudes, points of view, group thinking and other constraints making the cooperative working style difficult. Errors in decision-making are therefore essential contributors to design-induced erroneous behaviour of the system at the working level (cf. Bubb, 2002).

The link of cognitive aspects at the regulatory level At a first glance, regulation and the adjacent political level do not have an influence on safe operation of systems. Rather they appear as institutions assuring proper safety performance on the operational system. The distance to the working level, where the errors occur, is too high to perceive an immediate relationship. However, there is a mutual dependence of safety on the operational level and regulation. This dependence usually affects safety in the long term rather than the short term. Over-regulation or under-regulation may occur due to wrong judgements about the safety requirements that need overseeing. The fact that serious events may follow if regulatory oversight is mis-judging the safety relevance of a change was proven by a serious accident in the nuclear power plant Paks in Hungary in 2003, which resulted in a core melt event. As stated by the International Atomic Energy Agency (IAEA, 2003) the regulator had underestimated the safety-significance of a facility for cleaning fuel elements. Additionally, the management showed over-reliance on the safety performance of the manufacturer being accompanied by time pressure and limited human resources of the regulator at that time. These constraints led to the decision error not to perform a thorough safety evaluation of the cleaning system. The event is a typical example of under-regulation.

On the other hand, over-regulation may influence the human capabilities to recover successfully from a system failure. Too high standardization results in the reduced capability of the system to react adequately to disturbances (e.g. Grote, 1997). The personnel at the working level may be forced, for instance, to circumvent or violate procedures in order to recover a certain automation failure. Cumbersome procedures may lead to wrong decisions. Creative acts may be omitted in order to be compliant with the procedures. The analysis of events in nuclear power shows, for instance, that operators adhere better to procedures in disturbances (after an initiating event) although a more flexible and creative behaviour is required in these situations. On the other hand, adherence to procedures is low in standard situations, for which the procedures are designed (Sträter, 1997/2000). This contradictory effect can be explained by the users' fear of legal impairments as soon as they are 'under legal observation'. Wrong safety procedures therefore can lead to serious safety problems.

The regulatory level has a considerable role in preventing such problems by well-balanced safety procedures. To assure that regulatory requirements are well aligned to operational constraints needs a cooperative environment between both parties. In practice, gaps are developing between regulation and operation for several reasons like privatization of the industry, negative economic consequences of the industry if cooperating with the regulators, or similar. However, from the safety perspective, it is essential that no gaps develop between regulative requirements and operational constraints.

Cognitive Performance as a Common Cause of Accidents

Humans are forced to make judgements about the status of the system in a given context at all levels of operation. Increasingly, so-called errors of commission (EOCs) are observed in incidents and accidents. The term EOC subsumes those interventions of the user that are not required from the system point of view and aggravate the scenario evolution. The mid-air collision at Lake Constance 2002 is such an example, where nothing would have happened if the controller had not intervened in the automatic procedure (BFU, 2004).

Such incidents show the human in a Catch 22 situation (Hollnagel & Amalberti, 2001). The user has to judge the status and availability of a technical system, which may lead to a misinterpretation of the system-states under certain conditions (context). The user needs to make judgements about the system state and its validity in the particular situation. The user may then be wrongly blamed in two cases:

- If one fails to notice the automatic function is out of service and should intervene (usually called error of omission).
- If one wrongly perceives the automated system as not functioning and acts according to his/her own understanding of the situation, although the automatic system is working properly (usually called error of commission).

The nuclear industry, as one of the technical systems with the highest level of automation, has proven that the first case stems to, a great extent, from latent errors (Reason, 1990). Balfanz et al. (2004) show the impact of planning on the safety culture at the operational level. Problems of safety culture were shown as being induced into the system by incorrect maintenance or design of systems. The second case follows from the uncertainty of the system status induced by latent errors but also from inappropriate design of working procedures. EOCs are often to be observed in such situations. Violation of rules may be the most drastic consequence of such a system design (Shell, 2002).

Similar judgement processes can be observed in the management and organization of systems. For instance, one has to judge on the working level whether a certain procedure holds for a particular situational condition but the management decides on an inappropriate re-organization because it thinks the current organization is not working properly. As discussed above, one can also observe such judgements on the regulatory level. Regulators may over- or under-estimate the importance of a system change for safety.

All these experiences indicate that success or failure of the implementation, operation or maintenance of new systems depend heavily on the success or failure of the cognition of humans at those levels. In addition, the organizational level depends on the cognitive aspects. It is still humans who are acting at the organizational level. Therefore, cognitive and organizational aspects will not regarded independent or separate issues of safety in this book.

Accidents have to be seen as unwanted events, which have evolved over time and over all operational levels. Reason (1990) developed the famous cheese model to visualize the multiple contributors and causes in breaking the safety barriers. The defence in depth concept or barrier concept was introduced to stop developing deficiencies in a system into an accident (Figure 1.2).

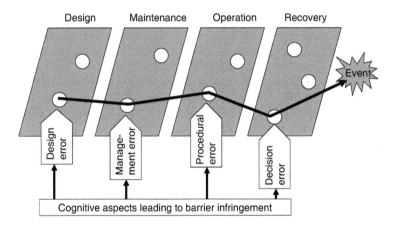

Figure 1.3 **Cognitive properties as a common cause for degradation of system safety barriers**

Referring to this visualization, human properties in information processing can be seen as the common causes generating unsafe situations at all operational levels (creating holes). Understanding cognition is therefore essential to understanding and preventing accidents. Similar mechanisms in human information processing are triggered at different operational levels with specific constraints.

Cognitive aspects explain why accidents continue to happen, but also how to recover from critical situations successfully. Cognitive performance generates and closes the holes in the cheese. Human practical decisions and constraints are therefore essential to understand and predict accidents. However, cognitive performance can also be very efficient, even in highly demanding and critical situations (GIHRE, 2004).

The cheese model is somewhat of a high-level safety model of a system similar to the event- and fault-tree approach. The development along the safety barriers may be understood as the event-tree, whereas the causes for the holes may be represented in a fault-tree approach. Human behavioural aspects are part of the event- and fault-tree. The link of human behaviour via cognitive processing to constraints therefore provides therefore a proposal to include organizational issues into Human Reliability Assessment (HRA) and Probabilistic Safety Assessment (PSA). However, the inclusion of these aspects into HRA-Methods and PSA models requires re-consideration of the fault-tree/event-tree approaches. As discussed in detail in Part III, the current way to assess human performance in PSA/HRA becomes complicated in modelling the impact of cognitive errors with respect to the following aspects.

- Inappropriate cognitive design will lead to critical situations in severe accidents since the high cognitive demands may lead to wrong human behaviour.
- Assessment of maintenance issues (low-power and shut-down modes in the nuclear industry for instance) is related to certain cognitive demands induced by the lack of training and procedures or the high variability of possible system states.
- Assessment of computerized control rooms is accompanied by increased cognitive activities (no more visual-driven information collection). Related man-machine interfaces will increase the cognitive demands of human process control.

A Proactive and Integrated View on Dealing with Cognitive Issues

The Hindsight Nature of Human Factors

Usually, the Human Factor (HF) community attempts to overcome potential problems in a product by investigating relevant HF-related issues in the design of systems. Workload and situational awareness are such concepts in HF research, as well as efficiency measured by reaction times. Examples are:

- In air traffic management, a pretty mature prototype of the system has to be in place to judge situation awareness or workload issues involved in the system (Sträter et al., 2003).
- In the car industry, navigation systems need to be maturely developed before distractions from the car-driving task can be assessed (Schweigert, 1999).
- In software, the Human-Machine Interface (HMI) has to be established in a functional prototype before acceptance tests can be performed (ISO-9241, 1993; Fukuda, 2004).

The incidents and accidents provided as examples above clearly indicate that this hindsight approach of cognitive modelling is not sufficient from the safety point of view. Often, the front-end of system operation is addressed as erroneous but deeper analysis of the causes reveals design issues related to the observed cognitive failures. Nevertheless, all the classical HF-methods are designed as measurements for the working level and come too late to make substantial system changes because the system is already settled. Design-related constraints, which will lead to serious safety problems in operation later on, will therefore not be covered by hindsight measurements (Balfanz et al., 2002). A proactive approach on how to tackle and prevent incidents is necessary. One effect of poor design and inefficient hindsight consideration of Human Factors is known from the introduction of new systems. The user is reluctant to accept new systems immediately and to build up trust in the system (e.g. in the Air Traffic Management environment; see Bonini, 2004 or Skyway, 2002). This reluctance may be reinforced by previous bad experiences. The trust issue links to another group of issues. Management decisions and organizational procedures are also defined in hindsight by using classical retroactive HF-methods. Again, the final system design is faced with constraints due to late consideration of managerial and organizational aspects on the working level.

Usually, methods such as rapid prototyping were developed to overcome the retroactive nature of HF-methods. Potential user-groups are faced with a mock-up of the new system. Their performance is then measured according to constructs such as workload or situational awareness. Prototyping means reducing the real functionality and real interaction with the system, and therefore needs a cognitive model to support and facilitate the process of reducing the reality to the prototyping. The same holds for Virtual Reality (VR) settings where only those aspects of reality can be represented that can be realized in the VR environment. However, the cognitive model is implicitly built into the prototyping process making prototyping a trial and error exercise rather than an engineering approach to cognitive design.

The use of hindsight approaches has an impact on the available budget for considering human issues. Usually, the money is already committed early in the development cycle and there is little flexibility in reacting to the HF issues left (Figure 1.3). Without an integrated proactive approach to HF aspects, these aspects may always be low cost-fixes in later stages of the life-cycle.

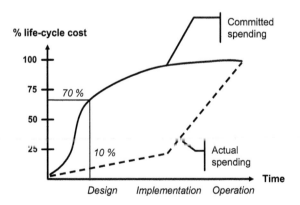

Figure 1.4   Project costs commitment in different product stages

Using a proactive approach would also enable the designer to see which cognitive processes and functions are affected by the system's design. This enables earlier assessment of design options and/or comparison of different systems (Sträter et al., 2004a). It also makes the identification of training needs more explicit and detailed. In addition, a proactive approach can be made compatible with incident analysis tools (Isaac et al., 2004). Hence, a proactive approach can be fed with existing incident data so that an assessment of potential errors can be made. Herewith, a designer may get an insight about what types of errors on the working level might be introduced by a proposed design under real contextual conditions. The approach therefore aims to prevent design errors and to increase the awareness of design about limitations at the operational level.

The ability to identify design issues facilitates the prioritization of HF measurements, such as *Mental Workload* or *Situation Awareness*, and can aid design decisions and simulation planning. Being able to measure skill changes as a result of design measures early (i.e. in time to influence design options and cost-benefit trade-offs) can also provide the opportunity for selection, training needs assessment and successful transition training as part of a more proactive technology change management.

Why is a More Coherent Picture of Cognition Needed?

One may argue that each industry and each problem should have its own cognitive models and that coherent cognitive models are of less importance. The potential user of the model could pick up a model that suits his problem best and leave the rest of models aside. Nevertheless, there are good reasons not to be satisfied with such a non-integrated approach to cognitive science.

Cognitive science to support system safety In several instances, hindsight-methods cannot be applied because the direct access or contact to the 'piece of interest' (the human being, or the group of persons, whose cognitive performance is being investigated) is missing. None of the usual cognitive design methods can be applied with sufficient validity. Among these settings are (Sträter & Bubb, 2003):

- Assessment of cognitive performance concerning humans in events or incidents based on evaluation of past-experience, where the access to the person(s) directly involved in the events or incidents is lacking.
- Assessment of cognitive performance in early design stages where an experimental setting has to be prepared, where a user group is not specified or a detailed specification of the design of the technical system is not available, so that none of the usual techniques can be applied.
- Assessment of cognitive performance in safety assessments where an unlikely or hypothetical situation of failures in the system has to be assessed, which never happened and will probably never occur.

These settings are typical application-areas for the remote prediction of cognitive processes and have a high interest in explaining and predicting cognitive aspects in order to achieve a safety design or organization. They have in common the requirement to perform a cognitive task analysis without direct access to a certain user or user-group. No direct measurements of human behaviour can be performed. Consequently, other data-sources have to be exploited to generate predictions about human performance and to derive design suggestions. In addition, such cases require a model that is robust enough to work in remote settings and allows description and prediction of cognitive processes to a certain level of detail. Those specific models are usually called Human Reliability Assessment (HRA) methods.

Cognitive science to assure safe systems Supporting the remote settings of system safety where hindsight-methods are not applicable is only one reason for a proactive integrated cognitive model from the safety point of view. As outlined above, cognitive aspects are leading to the holes in the safety barriers. Therefore, cognitive models are required for assuring that safety assessments are as complete as possible regarding human aspects in order to prevent accidents as far as possible.

Usually, we develop a model to structure our thoughts and to define the scope of what is to be considered in the problem space. This might be a specific design question (like development of a certain piece of automation) or it might be the proof of how a system works in a real environment. For all these issues, we usually have a certain cognitive model in our minds. Sometimes very explicit (as in HRA) but cognitive models may also be very superficial in the case of less structured approaches.

From the safety point of view, inhomogeneous models imply potential safety holes and hazards. The choice of a particular cognitive model for investigating safety issues implies that the safety assessment is only complete if the cognitive model is complete as well. In turn, an arbitrary collection of cognitive models leads to the conclusion that a safety assessment of a system may have considerable mis-assessments of safety in one or the other respect. If we have, for instance, a human error model that does not consider situational awareness or workload, this model remains incomplete regarding these concepts and is hence insufficient in its prediction of potential safety problems.

The examples of disasters mentioned above show that these cognitive aspects are, on the other hand the ones deteriorating the safety of the systems. Humans have the potential to overcome or circumvent technical means for safety (e.g. the technical interlocks of the reactor protection system were circumvented in Chernobyl, because the operational staff wanted to accomplish a certain task)

The problem of assessing cognitive issues is that we limit ourselves to those cognitive issues that are reflected in the model we use for investigation or assessment. Theis (2002) investigated the effect of incomplete cognitive modelling on safety in the framework of decision errors during car driving. She showed in particular that there is a decisive difference between purely task-related cognitive modelling and context-related cognitive modelling, which includes the goals and intentions of the car driver. The risk involved within a system may be mis-assessed by the factor of 200 if a safety issue is approached from the perspective of a certain cognitive model that neglects the relevant cognitive mechanisms triggered by a certain context. The results are highly relevant for Human Reliability Assessment (HRA), where the decision-based errors are still not included in safety assessments, although they have proven to be the ones leading systems into disasters. Safety assessments therefore are an important customer of a homogeneous cognitive modelling in order to be informed about the safety status induced by human cognitive processes.

The performance of cognitive science within the spectrum of disciplines in system design Other customers might be system designers. They also need to have an integrated picture of the cognitive issues they imply into a certain system. If a system is designed according to minimization of workload for instance, it is not necessarily reflecting an optimal system from the perspective of error prevention. Designers, usually not, or little, trained in cognitive science, would regard both as independent concepts and would make independent assessments, which may lead to neglecting one or other issue in the final design.

A typical example for independent treatment is the problem of hybrid systems (i.e. systems with a mix of old and new technology). Hybrid systems evolve from the introduction of new systems into existing ones. The dependencies of workload, when switching between both technologies, are often neglected, or at least under-estimated, during the design of the new system. New systems are often developed as if being operated as an isolated new system.

In addition, training specialists, who have to prepare the staff in working with new technology, may require an integration of all cognitive aspects, because they need to train the cognitive skills, knowledge and attitudes of staff independently of whichever cognitive concept is used for designing a system.

Modelling means reducing the richness of the world in such a way that the reduced view serves a certain purpose (Hollnagel, 1999). It is therefore not the aim of cognitive modelling to describe the entire human as a *Homunculus* or electronic human being. It is rather more a summarizing perspective on the principle cognitive behaviour that has to be considered in an appropriate system design and that allows predicting behaviour within a particular situation. Naturally, such a procedure of classifying the world has the disadvantage (and advantage) to loose information. It will never be able to predict all possible futures of a certain cognitive behaviour. A proactive modelling perspective will therefore never be able to substitute HF measurements but it will be able to focus the experimental setting onto relevant questions, to provide input in preparation and pre-training of experimental participants and to reduce the financial effort of experiments. Relevant questions to be answered by integrated cognitive modelling in design are (adapted from Sträter et al., 2003 and Bunting, 2003):

- How much will workload be affected by the designed task and unspecified additional tasks (including coordination effort for following independently designed tasks)?
- How long will a certain action (related to a specified task performance) last?
- What will be the effect if a human cooperates with automation?
- How will performance be impacted by the layout of the control system?
- How will a team perform in complex and/or remote settings?
- How will environmental aspects influence performance?
- How will functions be allocated optimally?
- What will be the workload of the operators?
- What will be the situational awareness of the operators?
- Which skills will be needed to perform successfully with a system?
- How probable will be success or failure of humans within a certain system?
- How reliable will be the human behaviour?
- How much variability will be there in the way a human performs a certain task or how much is the functional variance respectively?
- How much will the performance degrade under certain environmental and contextual conditions?
- Will the user trust the system?

Cognitive science applied as a tool for system management Any management has to be aware about the interrelations of cognitive issues for getting the organizational aspects of a system right. Too often, only those HF concepts of cognition were considered in reorganizations that were currently 'en vogue'. The design of the systems regarding other HF concepts was neither investigated nor considered. The system then broke down due to the non-considered HF aspects. A typical example is the introduction of a new system promising the reduction of workload but neglecting the phase of transition from the old to the new system and neglecting that the new system has to be operated together with other, old equipment. Without preparation of the staff, the workload increases during the transition phase and the efficiency of the overall system decreases. Similar effects arise from system malfunctions or disturbances. Software development is full of such stories and, consequently, ISO-9241, the ISO standard on software ergonomics, states that participation of staff and operational contexts are essential criteria for the development and introduction of software products (ISO-9241 Part 10, 1993). The same effects were observed in the introduction of new operational rooms or new airspace layouts in Air Traffic Management, hybrid control rooms in nuclear industry, navigation systems in ground-transportation, or the issue of mental workload in occupational safety as stated in the ISO-10075 (2002).

Operational managers, safety managers, regulators and operational staff concerned with human factors and safety issues should be aware of the interrelations of cognitive aspects. Such awareness cannot be created if the cognitive models stay independent with unclear interfaces between them. A common model also facilitates learning and understanding between the different operational levels, regarding their cognitive behaviour and decisions.

Cognitive science as a business tool   Cognitive science cannot be applied independently from other areas of system design and system improvement. In order to get appropriate resources in the product development, maintenance and improvement, the cognitive issues have to compete in terms of economic relevance with other domains like software or hardware development, or investments in automation. Long- as well as short-term relevance and benefits of means in the cognitive areas compared with other means for system development or improvement have to be known. Cognitive science would need some approach for a homogeneous picture considering the different cognitive aspects and their interrelation.

Many cognitive scientists would counter-argue that the cognitive area is far too complex to generate some kind of coherent picture. On the other hand, a non-integrated picture of cognitive aspects would make cognitive science some kind of supportive service but never a discipline of product design. Other fields of business have means to derive indications based on integrated models of their world although their world is complex as well. The best-known example is the stock market where a single value is generated although the mechanisms behind are also complex and difficult. Certainly, an integrated picture of cognitive aspects would not mean having a single performance indicator like the stock indices.

An integrated picture of cognitive aspects would provide a picture on how the different approaches are structured and interrelated so that some general assessments can be made without neglecting the complexity of cognitive issues. Two aspects are of considerable importance for achieving such a business-related approach for cognitive science. Figure 1.4 illustrates this need.

Usually, the causal and contributing factors revealed from incidents show unwanted states in preceding system life-cycle states, such as poor system design or organization of basic work-procedures. To overcome such design factors as contributors to future events, one has to step back into the life-cycle stages. Strictly speaking, a design error occurred in an event during system operation can only be avoided by re-iterating the design stage and all the succeeding states. However, this is hardly done, because changing the design (and subsequently re-iterating the implementation phase) is cost and time intensive (Sträter & Van Damme, 2004).

After an event, it is tempting to look just for a countermeasure on the implementation or operational phase. The costs for means are comparably low and measures comparably easy to implement. Well-known means to prevent staff from doing certain actions are, for instance, revisions of procedures, a simple change of workplace arrangements, or the creation of technical interlocks. The efforts increase if the causal factor can be traced back into design or concept issues. Such factors will usually not be tackled, because the costs for implementing changes increase by the costs of the design or concept phase respectively.

The cost of implementing changes depends upon where the changes have to be made in the production cycle; changes in one product stage can influence and/or require additional changes in other elements of that stage. These may simply be additive, or, in the worst case, multiplicative. Therefore, important latent errors due to factors in design and concept are usually not addressed well enough. The precautions remain sub-optimal and hence they will not prevent the events from happening again. Examples of how little causes for events fit with the countermeasures undertaken are shown in Sträter (1997/2000, p. 210) for the nuclear industry.

Figure 1.5 also indicates the imbalance of available and required knowledge about cognitive issues. Prospective methods have to be compatible with the retrospective methods in order to generate efficient and robust recommendations regarding cognitive aspects throughout the life-cycle. To implement cognitive science in business, both parts have to go hand in hand. Such a proactive approach also allows linking decisions on the management level with HF issues on the working level. Moreover, it prevents accidents or incidents by better avoiding unwanted poor design and therefore prevents unwanted financial losses in the long term.

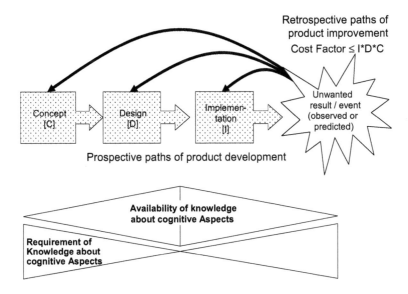

Figure 1.5   Human Factor elements in different product stages

Cognitive science to combine incident reporting and safety assessment methodologies Events show the importance of Human Factors for the safety of technical systems. In industries like Aviation and Air Traffic Management (ATM), the portion of Human Factor related events varies around 70% to 90%. ATM, with its considerable reliance on human interactions, belongs to the group of industries where human interventions lead immediately to safety issues because they are the final safety net.

Often Human Factor related events are directly connected to the term Human Error. Although, the term *Human Error* is widely used, it unfortunately implies that the cause of the event was due to an insufficient interaction of one or several persons with the technical system. However, in technical terms, Human Error comprises all elements of the technical and organizational conditions that led to a certain, insufficient interaction of a human with the system. The term Human Error therefore comprises all factors (internal and external) which may lead to misbehaviour of a human.

At a first glance, the simplest attempt to overcome human errors is taking the human as much as possible out of the loop by introducing automation. Experiences in highly automated industries, in particular in nuclear safety and aviation, have however shown that automation is instead shifting the problem of human errors into maintenance and design rather than eliminating them. With increasing complexity of systems, human errors spread out into all life-cycle phases, from concept development and prototyping via design and validation to implementation and operation. Human errors also spread into the different parties involved in assuring safety operation of systems (operation, management, regulation). Cognitive modelling is a basis to assure that a common understanding is developed throughout the life-cycle and throughout the involved parties on both retrospective and prospective safety assessments. Two essential elements have to be established:

- Feedback loops from succeeding to preceding design stages. They include recommendations for design strategies, for rapid prototyping, and for validating experiments in the product development phases and recommendations for event reporting schemes in the product implementation phases.
- Predictive elements for the appropriate design and implementation of technology, such as design decision support systems in the product development phases and Human Reliability Assessment (HRA) methods in the product implementation phases.

Cognitive science as a science   Finally, if there is no evidence of how different models can be represented in a homogeneous way, cognitive science has failed as a science. The hindsight nature of HF measurements conflicts with the desire of prediction of science. Scientific rules, which must hold for any research, are in particular:

- *Principle of simplicity of nature*: Nature is simple. Simplicity means that it is very unlikely that nature has created a cognitive apparatus that cannot be described and explained by relatively simple mechanisms. It is unlikely for instance that the functioning of the cognitive apparatus has any similarities to a computer structure (although this metaphor was taken some time ago to explain the difference between short-term and long-term memory; see a detailed discussion in Chapter 2). Nature would never have built such a complicated and error-prone system as a computer. Consequently, a theory about cognitive processes has to be simple as well, i.e. the basis and statements of such a theory have to be simple, logically built and easily traceable by observable behaviour and data.

- *Any model has to obey the rules of scientific logic*: Any model can never be verified but every model can always be falsified (Popper, 1997). This statement refers to a picture Popper has drawn: the observation of a set of white swans may lead to the expression 'All swans are white'. This empirically derived statement can never be verified, although the set of observed swans may become larger, because we will never be able to see all past, present or future swans in the world. However, the observation of only one black swan would clearly be sufficient to falsify this statement. Applied to cognitive modelling, the existing arbitrary collection of cognitive models enhances the danger that the models are not compatible with each other and will mutually falsify each other.
- *Any new model should include the explanatory power of the preceding ones*: Any new model has to be capable of explaining the aspects the former models were representing. Hence, a central requirement for model building is that any new or advanced model must not contradict the key findings of its preceding models. To put it another way, the integration of preceding models within an advanced model is already the first validation of the advanced model.
- *Any new model should have more explanatory power than the preceding ones*: Any new model should also represent new or additional findings or should be able to integrate a variety of findings.

Questions for Developing a Proactive and Integrated Approach

The most serious problem of cognitive modelling is the arbitrary or even incompatible collection of models. Often the concepts behind the models are reflecting 'a fashion' rather than scientific cognitive modelling. What is called a cognitive model often turns out to be a superficial description of a working process. This prevents cognitive modelling not only from being perceived as a science, but it also generates problems of support to system safety, of integration with other disciplines of system design, of application in system management and business planning. Consequently, a potential customer of cognitive science may become disappointed or at least confused by the number and the variety of cognitive models. There remain doubts, which model is suitable with respect to a certain situation or a certain problem.

As an example, workload models, situational awareness models or human error models are seen as rather separated approaches to deal with different HF problems. Human error models take workload as a Performance Shaping Factor (PSF), while workload models describe the resources and processes that are impacted by the workload. While in the former, workload is seen as a one-dimensional intervening variable, it is seen as a multi-dimensional concept in the latter. According to logic, there is no third ('tertium non datur'), which means that both perspectives must be coherent if the concept of workload is true and the concept of human error under workload is also true. All are dealing with cognition of humans and have therefore a common source, namely the human cognitive system. The human mind will not be different if one investigates errors, workload or any other cognitive aspect. However, a common ground of all the different models and concepts is hardly to be seen by a customer or even by an expert in cognitive science. According to the logical expression 'tertium non datur', there should be a more coherent picture of the common ground of all these models.

A common ground of cognitive modelling is necessary because, although cognitive models are used for system design and safety assessments, accidents continue to happen. Obviously, cognitive modelling is not able to provide an appropriate input into system design and error prevention. This situation could be due to

- the incompleteness or incorrectness of the cognitive model used, or
- a lack of using cognitive models in design and organization of systems, or
- a lack in applying cognitive models in safety assessments.

A proactive integrated approach for cognitive issues has to be developed from the following questions on human cognition:

- How can psychological and physiological aspects as well as mathematical capabilities of human cognition be modelled? This question will be elaborated in Part II of this book. A generic but sufficiently detailed model on cognition suitable to explain a range of HF issues, such as workload issues, decisions as well as human error modes, will be presented there.
- How can human cognition be integrated into ergonomic design and organizational processes? Solutions to this problem will be elaborated in Part III of this book. It will be described how problems of cognitive performance in operation can be prevented by considering cognitive aspects in the design phase and by recognizing cognitive aspects of team-working in the work organization. Moreover, Part III is also concerned with the application of the model of human cognitive processes in management. Errors or non-optimal decisions at the management level are known as important contributors to accidents and the sub-optimal design of technical systems regarding cognitive aspects.

The next chapter will discuss the reasons for, and an approach of how to address, these issues in cognitive modelling.

Chapter 2

The Genesis of Modelling Cognition in Safety

Modelling cognition in safety needs a balance between psychological soundness and applicability. On the one hand, it requires representing the cognitive processes that are safety relevant. On the other hand, it has to be practical enough to be used in engineering design or safety assessments. The trade-off leads to differences between the models as described in the scientific literature and the way they are practised. However, safety needs a valid approach that covers safety relevant human characteristics as much as possible. The high-level structures of current models in use for safety assessments lack certain human safety issues to a considerable extent. Therefore, the essential aspects of human cognition for safety need to be represented as thoroughly as possible on the high-level structure of a model in order to assure that safety relevant aspects are considered in practice. The genesis will show how these high-level structures were developed, where the current constraints of using models for safety come from, and what the missing aspects are.

The Early Psychological Disputes from Behaviourism to Cognitivism

Many of the current problems and limitations in cognitive modelling have historical reasons in the disputes and approaches developed during the last century. To understand the needs and problems in cognitive modelling of today, these are discussed briefly.

Behaviourism and Psychological Measurement

Behaviourism was founded with the observation of animals (Lorenz, 1978). Behaviourism is trying to make predictions about human information processing by investigating the relationship of input and output, as these are the observable parts of human behaviour. The active role of the information processing is excluded. Even now, the basic principles of behaviourism persist in all psychological measurements. A simple coherency of the input (stimulus S) and output (response R) is supposed and data are used to validate the hypothesized S-R relation. The coherency of input-output relationships is expressed as a dimension of human information processing (Figure 2.1).

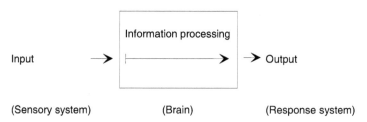

Figure 2.1   The behaviourism paradigm

As human information processing cannot be represented in only one dimension, the elementary approach depicted was extended into multi-dimensional concepts (batteries of indicators) and different aspects of a dimension were modelled independently (e.g. crystal and fluid intelligence).

The impact of behaviourism on cognitive modelling cannot be overestimated. The paradigm of behaviourism can be observed in virtually every cognitive model, in particular in models used for error analysis or error prediction. Furthermore, it can be found in the structure of questionnaires or methods used for analysing working conditions (e.g. Schimdtke, 1993) or in models for understanding stress (Hockey, 1984). This idea of understanding the human as an information processor, where information is processed from the perceptual system and categorized in order to generate a response, has never changed.

The idea of mapping signals and responses is also a basis in human error modelling, as suggested by Norman (1981) and Weimer (1931) later on to explain different types of cognitive human errors. They distinguish errors in information processing (so-called slips) and errors in applying schemes (so-called mistakes). However, the aspect of assimilation and accommodation was never systematically elaborated further as a potential part of modelling active human behaviour or errors of commission.

The active part of human intelligence, the creative act and creative decision-making never found a suitable integration into this information processing perspective. The discussion and accidents described in the introduction proves the behaviourism assumption to treat humans as more or less passive information processors to be false. The current problem of errors of commission in highly automated systems cannot be tackled with the information processing perspective.

The Discovery of the 'Mystical' Aspects of Active Human Behaviour

The discovery of 'mystical', non-rational human behaviour showed the active parts of human behaviour and that the human information processing approach is insufficient. In particular, the extreme cases of mental diseases (like schizophrenia) make obvious that a stimulus-response paradigm is far from sufficient to understand the nature of human information processing. Freud (1930) was the first major scientist to approach the active part of human behaviour by distinguishing the controlled, rational or conscious information processing from the uncontrolled, irrational or unconscious part of the human information processing nature. This aspect was re-investigated by Schneider and Shiffrin (1977) later on in cognitive modelling and led to the important distinction of sequential and parallel processing.

The first classification of the active nature of human information processing was the collection of the so-called *Gestalt-Laws* (e.g. Guski, 1989). Perception is largely drawn by higher-level mental processes completing the given information. The Gestalt-Laws show clearly that knowledge-based characteristics, such as decision-making, expectations and intentions, are not excluded from skill-based perceptual processes.

Research on learning showed the importance of memory for understanding active human behaviour. By investigating children's way of learning, Piaget (1947) identified two general mechanisms memory that operates in: assimilation and accommodation. Assimilation describes the acquisition of new information and processing according to stored rules. Accommodation describes the process of generating new behaviour and skills by the integration of acquired information into the existing knowledge or schemata. DeKleer & Brown (1983) transferred Piaget's concept of learning into cognitive science. They distinguished the envisioning (acquisition of a mental model) and the running of a mental model. Johnson-Laird (1983) suggested using the concept of mental models as the basis of learning and active human behaviour and integrated the propositional inference logic into the mental model approach. Moray (1987) integrated the memory aspect into mental models by introducing so-called lattices. Here, memory is understood as linkages between signal and responses.

Cognitivism and Technomorphism

The rapid technological development in the time after the Second World War had a huge impact on cognitive modelling and many of the 'mystical', active aspects of human behaviour were getting out of the focus of cognitive modelling. During the Second World War, people had to function according to the information received and there was no room for modelling or understanding the active or creative part of human behaviour. It was also too tempting to align psychological modelling with technological modelling and to make the psychological models as successful as the technological ones. Models stemming from the desire to equalize with technology may be described as techno-morphologic models. They can simply be mapped to their contemporary technological developments. Nevertheless, they also show certain cognitive aspects that are of importance to be considered in an integrated approach.

Analogue representation and mental models If one describes the way to travel from town A to town B to town C from memory, the time it takes to describe the ways between the towns is proportional to the time it takes to follow the roads on a map with a finger. It appears as if an 'inner eye' is used to scan the memorized map in the same way as the external eye is scanning the real map. Kosslyn (1973) revealed the idea of analogue representation and an inner eye that performs the cognitive activities from this finding. The model of the 'inner eye' was inspired by the technological innovation of transmitting pictures remotely via radio transmission. Analogue representation means that the proportions of the real scene are kept between the real and transmitted virtual picture. Finally, this development resulted into the construct of mental models (e.g. Johnson-Laird, 1983) and the *Pandemonium*-models, representing coordinated agents in the brain performing certain mental models (Rummelhart et al., 1972).

Resource sharing and filter models   Experimental psychology observed interference if different information has to be processed at the same time. The dual task paradigm was developed in order to investigate this effect in more detail (discussions in GIHRE, 2004). In dichotic listening, where the left and the right ear are exposed to two different streams of auditive information, it is observed that human can only process one of the information streams. Furthermore, the words used in one stream may be mixed up with the words of the second stream (Jones et al., 2003). Interference may also occur between different modalities (auditive and visual processing, Wickens & Hollands, 2000). The question is how to integrate the different findings on interference and the limitations in processing a certain amount of information. An early model was developed by Broadbent (1958). He developed a filter-model to explain the findings of the interference effects inspired by the development of electronic filter systems (as used in Radios, TVs or radar systems etc). Alternatively, Wickens suggested explaining the interference effects as a result of limited resources that either have to be allocated to one or the other information. Humans may have a fixed amount of a certain 'fluid' available to steer their attention to certain information (focused attention in performing a certain action for instance) or to spread attention between several (divided attention in monitoring tasks). Other resource models exist for the concepts of stress, emotion, motivation, pain, or fatigue (see Hockey, 1984 for instance). Both metaphors (filter and resource models) are borrowed from the state of technology at the time the concepts were developed. In order to compensate for the complexity of findings in a range of follow-up experiments, these models were extended again into the direction of multiple filter models or multiple resource theory (MRT). However, the general approach of the concepts was never revised.

Control models   McRuer (1967) revealed in a couple of experiments that a human being is able to adapt himself to many dynamics of a system and is able to perform almost as an electronic control system. The control model is very suitable to describe the dynamics of man-machine systems and is used successfully in aircraft and car design. However, the model only allows modelling highly skilled perception-driven actions and is not able to model higher cognitive functions (Bubb, 1993).

Information theory   Probably the most important techno-morphologic model is the information theory in cognitive psychology. Information theory was originally developed to describe the information transmission in radio-technology or code development but was quickly transferred to psychology (Attneave, 1974). Miller (1956) showed that the capacity for repeating randomized units of information (chunks) is limited to 7±2 chunks at a time. Further investigations showed that this result could be described using the mathematical approach of the information theory. Later on it was shown that this limitation is also dependent on the dynamics of the tasks. The higher the dynamics the fewer chunks can be distinguished at a time (see discussions in Wickens & Hollands, 2000; Bubb & Sträter, 2005). In very dynamic systems, this may even decrease to one chunk per time. The time the number of chunks can be kept in the memory also depends on the number of chunks to remember. While seven chunks can only be recalled for about 7 seconds, one chunk can be held in the memory for minutes (Card et al., 1983). The effect of the limitation was first named *Memory Span* (Katzenberger, 1967). However, it is currently called short-term memory (STM), which is a result of the next techno-morphologic approach, the computer metaphor.

The central processor and computer metaphor A further step in modelling cognition was accompanied by the growing advantages of computers and information transmission. Models of this type are widely summarized under the heading *Computer Metaphors*. These models try to explain some mental mechanisms by choosing the computer as an example for a human brain. The computer metaphor is probably the most tempting and most problematic metaphor ever used to describe cognitive processing. Computers are, on the one hand, information processors and many of the aspects of information processing can be simulated or at least demonstrated with computers. On the other hand, the metaphor is the most dangerous because it provides a wrong picture of how cognition is taking place. The most problematic computer metaphor is the distinction between short-term memory (STM) and long-term memory (LTM).

In a computer system, both are distinct types of storage devices (random access memory vs. hard-discs). This picture was transferred to cognitive modelling in order to integrate the findings on the memory span with the fact that we can memorize more than just seven chunks in our memory. This metaphor was exploited far too much and only recently is the distinction between STM and LTM being discussed as a wrong assumption made over the last few years and that both are the effect of one common architecture of memory (cf. Jones et al., 2003). However, the understanding of a common architecture of memory still needs to be defended widely against the classical view of two independent types of memory, although it has the advantage of making cognitive modelling more valid and more understandable than a model learned from the structure of micro computers. Opposite of the common knowledge that humans do not have computer chips in their brains, this idea is still very vivid. It was too easy to sell the idea during the computer revolution, although the original term of memory span would have made better progress in finding an appropriate model of memory that is less misleading in the sense that there are two different physical entities (as there are in computers).

A consequent further development was the assumption of a central processor that uses the information stored in STM and LTM and hence is performing the cognitive activities. Similar concerns as for long- and short-memory hold for the central processor metaphor. Again, the cognitive model about the central processor is borrowed from computer architecture. One could have easily seen that this metaphor could never validly hold for human cognitive processing if one ever had tried to break this metaphor down to the level of the micro-programs of a central processing unit (CPU) of a computer. The micro-program is hard-wired software that is needed in any processor to link the hardware to the higher-level software (as assembler, c or similar). The cognitive version of a central processor never reached a corresponding level of detail.

The software metaphor from goal notations to object orientation   Since any hardware needs some software, finally the variety of software metaphors rounded up the 'human being as a computer' analogy. The (very) important distinction between conscious and unconscious processing was explained as being similar to the distinctions between interpreter and compiler (Anderson, 1996). Consciously controlled processing was seen as being based on interpreted knowledge whereas unconscious, automated processing was regarded as being based on compiled knowledge. Besides the analogy between compiler and interpreter, a couple of programming languages were developed such as Lisp or Prolog, which attempted to reflect the architecture of the human memory.

Varieties of systems were developed to mirror the cognitive aspects in computer languages. Expert system approaches, database approaches or knowledge management approaches belong to this category. Most famous developments in this area are the GOMS model (Goal Operations Methods) and Selection Rules Model and the related cognitive complexity theory (CCT) of Kieras & Polson (1985), the adaptive control of thought approach (ACT, Anderson, 1983), or the PSI model (Dörner, 1999).

An important aspect introduced by this type of techno-morphologic approach is the role of goals and internal processing states for the decision-making and resulting behaviour. Software ergonomics took the importance of goals into account. To match the users' needs is one of the most important criteria for software evaluation (ISO-9241, 1993). However, the architecture of the software system is far from being able to represent the architecture of the system intended to model (i.e., the brain). Other models were developed that use the computer as a computational tool but are more active of the physiological architecture of the brain, the so-called neurological or connectionism approaches (Kempke, 1988). The unified models of cognition network SOAR attempts to integrate these physiology-related models with the models stemming from the software-analogy (Waldrop, 1988).

The Gap between Issues and Models

This brief overview has shown how much cognitive modelling is linked to the state of technological development as well as to the general technological and social thinking of the times in which they were founded. Many of the concepts received revisions or refinements instead of a complete reconsideration. Significant disputes between the different approaches occurred instead of an attempt to integrate them. Sommer et al. (1998) and Sanders (1983) describe the dispute between resource theoretical approaches and filter approaches in attentional research. Jones et al.

(2003) describes the dispute over short and long-term memory as being of different nature or being based on the same neurological structures. However, all are observations from the same origin (the brain) and the dispute shows more a discussion between competing products than a scientific process of integrating the different aspects of one and the same physical object.

It is pretty obvious from the perspective of today's knowledge about cognition that techno-morphologic approaches of modelling are a rather desperate attempt to sneak ideas from technology, with the hope of finding an approach to model the cognitive apparatus. As we know now, many of the assumptions and analogies made cannot persist. Techno-morphologic pictures are good for generating ideas or describing experimental effects, but they are already used as models in cognitive science far too often. However, all the different approaches have some interesting aspects to offer. Thus, many cognitive approaches we currently use in design or error prevention are still based on the metaphors established in this techno-morphologic age of cognitive modelling. The mostly used human error prediction method THERP (Technique for Human Error Rate Prediction, see Swain & Guttmann, 1883) is founded on the classical behaviourism approach, for instance.

Despite all the difficulties in taking technological developments as a source for generating a cognitive model, all metaphors have some truth. The findings they represent are important aspects of cognitive information processing. Three aspects should be emphasized.

- Analogue representation approaches show the importance of the active or creative aspect of human cognition. This considerable aspect of self-perception and reflection (called the 'inner eye') was never treated sufficiently in cognitive modelling. Self-perception and reflection are essential aspects of cognition because they are vital to understand human capabilities to generate ideas, to monitor one's own thoughts and to be creative. Philosophy considers the reflective aspect of cognition as an essential aspect for distinguishing humans from animals. Animals' abilities to recognize themselves and their behaviour are very limited. Descartes therefore made the famous statement about the nature of human cognition 'Cogito ergo sum [I think therefore I am]' (cf. Störig, 1988). The accidents discussed above show that this active part of cognition could be much better integrated in the models used for analysis and assessment of cognition in order to prevent such accidents. Humans circumvent procedures and intervene into the reactor-safety protection system because they are confident of doing the right actions, or they ignore alarms from the Ground Proximity Warning System (GPWS) because they are convinced that the plane is at the right altitude. Such human errors cannot be explained by assuming a purely information processing related human cognitive behaviour.
- Control and filter models show the importance of time in human information processing. Even complex tasks, like flying aircrafts or driving cars, are incredibly fast and well aligned to the system being operated. However, this view of cognitive modelling was never integrated into the models that

concentrate on decision-making. Control models are seen as a separate approach to model human behaviour in dynamics. However, in tasks where both aspects (decision-making and dynamics), come together, integration is needed. Currently this is the case in aviation, air traffic management and car driving, where automation is introduced into dynamic systems. But also, decision-making of medical doctors during surgeries can only be understood if the time aspects and the decision aspects are investigated in a combined manner (GIHRE, 2004).

• The computer metaphor reveals the importance of considering the computational or mathematical feasibility of the model generated. The requirement of computational or mathematical effectiveness can serve as an additional means to find out whether the structure of the model fits with the structure of the piece of interest, the human brain.

The techno-morphologic age of cognitive modelling has led to two important general models with high relevance for the current situation in error analysis and prediction: the stage-based approach and the phase-based approach of human information processing.

The Seduction and Constraint of Information Processing Stages

The first attempt to integrate the variety of findings for an engineering use of cognitive modelling was defining stages of information processing. This approach constitutes an obvious and easy-to-follow approach of human information processing at first glance. However, the stages defined do cause most of the problems in applying cognitive processing models to system safety. How does this fit together?

The stages of cognitive information processing are defined based on the paradigm of input (stimulus) and output (response). Input and output are considered as interfaces to an environment that has rather little impact on the method of human information processing. Information processing is broken down into different stages, like perception, decision-making execution with input from memory and is steered by attentional resources that are spread over the functional model. Figure 2.2 shows the scheme of Wickens for human information processing as a classical approach of integrating various cognitive findings. However, this figure should also hold as an example for a range of approaches for modelling cognition as information processing stages. Psychological literature has plenty of these types of models as attempts to model cognition. All such models are based on the behaviourism and techno-morphologic metaphors, in particular the resource-model and computer-analogy.

The definition of stages is rather arbitrary. Wickens defined three stages and two storages. Other approaches use different numbers. According to Wickens (1984) one can also find four stages defined (encoding, recognition, response selection, execution). The THERP model (Swain & Guttmann, 1983) defined three stages (perception, cognition, execution), Rouse & Rouse (1983) defined five stages (detection, comparison and decision, response selection, action selection, action feedback), and Rasmussen (1986) defined eight stages (activate, observe, identify, interpret, evaluate, define task, formulate procedure, execute). The discrepancy in the number of stages gives reasons for considerations from the safety perspective because the number of potential errors may vary based on the model. As a consequence, the number of prevention means may vary as well. The reason for the fuzziness of the number of information processing stages lies in the way the stages are experimentally generated.

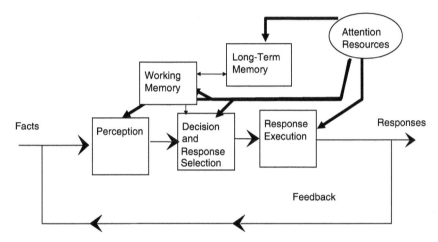

Figure 2.2   Wickens scheme of factors for human information processing; a classical approach of integrating various cognitive findings (according to Wickens, 1984)

The Inquisitory Logic of Modelling Information Processing Stages

Stage-based approaches are founded on the Sternberg paradigm or the so-called *Additive Factor Logic* (Sternberg, 1969; Anderson, 1996; Wickens, 1984). The additive factor logic is a fairly straightforward way of how to come from measured observations to models. Figure 2.3 illustrates this process. In an experiment, a set of numbers needs to be memorized and a particular button has to be pressed if the number presented on a screen belongs to the memorized set of numbers. The reaction time from stimulus presentation to pressing the button is measured. The quality of the stimulus is varied (e.g. parts of the stimulus are hidden vs. stimulus is shown entirely). It will be observed that the reaction time increases if the stimulus quality worsens.

Now this basic experiment has two variations. (A) The set of numbers is varied between low and high. The response time is then measured according to the tasks under good or bad stimulus quality. The reaction times measured for this variation may result in a graph like Figure 2.3A. (B) The set of numbers is varied between a set of distinctive numbers and a set of numbers, which are easily to confuse. The reaction times measured for this variation may result in a graph like Figure 2.3B. The additive factor logic classifies possible variations into independent or interdependent variations. Independent observations show an additive effect of the reaction time and therefore lead to the conclusion that two independent processing stages are needed to explain the difference between reaction times. In the case of interdependent observations, these obviously stem from one processing stage.

Further investigations were undertaken to underpin the concept of the additive-factor logic. Amongst the variations are investigations of the stimulus quality, stimulus discriminability, compatibility, expectancy and complexity. The more studies performed according to the inquisitory logic of the approach, the more stages were identified and the more the approach showed that the sequential information-processing model is unable to represent the observations (cf. Wickens, 1984, p. 339).

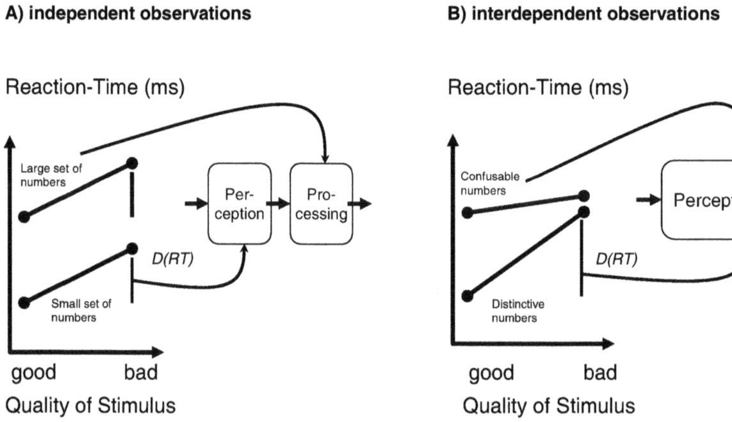

A) independent observations   **B) interdependent observations**

Figure 2.3   **The additive-factor logic or Sternberg paradigm**

The additive factor logic obeys an *Inquisitor's Logic*. Any variation of any parameter has only two possibilities: to be classified either as being an influence on a certain stage or being a new information processing stage. This logic is similar to the one used during the medieval inquisition era. If a woman was accused to be a witch, one proof was to tie her up and to throw her into a river. If she was able to unleash and rescue herself, she was proven to have dark forces and to be a witch. Consequently, she was condemned and burned at the stake afterwards. If she was not able to unleash herself, the accusation was wrong and she was apparently innocent. In both cases, the outcome for the woman was the same.

Problems of Stage-based Approaches

Misleading error prevention   The stage-based approach shows, like no other approach, how much our error prevention and thinking on safety is related to the model we assume of human behaviour. The sequential models persist as models to structure human information processing. They are easy to understand and follow and they have the advantage of being of use in an engineering sense. Arguments to abandon such models for modelling human error are manifold (e.g., Hollnagel & Amalberti, 2001). However, they have not managed to change the general concept. The historical grounds and the techno-morphologic intuitiveness are too firm. They also appear suitable in particular for describing human errors at a first glance. According to the cognitive model outlined in Figure 2.2 (above), possible cognitive reasons for an inappropriate response could be:

- wrong or inappropriate execution;
- wrong or inappropriate decision-making;
- wrong or inappropriate perception;
- wrong or inappropriate use of working memory;
- wrong or inappropriate use of long-term memory;
- wrong or inappropriate allocation of attentional resources.

However, although such structuring of human errors is easily understandable, it is dangerous in applying the structure to safety. It limits the causes for human errors to the human being and does not distinguish between the phenotype vs. genotype of human error (e.g. Hollnagel 1993; Bubb, 1992; Sträter, 1997/2000). It limits the potential causes to the human being and consequently limits the potential countermeasures to humans as well. Triggering aspects, like inappropriate design of the stimulus quality or misleading indications from the context, are just taken as intervening variables on the cognitive processes but not as aspects influencing the sequence of the cognitive processing.

Such an approach also steers our thinking about potential measurements and prevention means for improving safety or reducing human error impact. For reducing the inappropriate allocation of attentional resources, we try to increase the situational awareness of the persons (Endsley, 1995) and have a resource model in our mind while doing so. To limit errors in perception, decision-making and execution, we reduce workload or develop decision aids and have a filter model in our mind while doing so. Paper-based or electronic procedures should support our long-term memory. Training concepts should overcome inappropriate use of short-term memory. Whether all the means lead to improved safety is dependent on basic models we have regarding cognitive information processing.

Finally, the use of such simplified cognitive models may lead to additional safety problems because the means produce procedural overruling or scattered displays. In about 43% of all nuclear events with human factor involvement, training is suggested as a countermeasure to overcome the problems (Siemens, 2002). Such measures reflect the information-processing model we assume for the human involvement in the system rather than being effective counter-measures to prevent human error and to assure safety.

Lack of validity of the stage-based models   The stage-based approach is a historically grown construct of human information processing modelling. As Anderson (1996) states, the model was generated based on the computer metaphor, which was the leading metaphor in those days. He further states that the logic was invented according to the technological possibilities of those times to perform experiments. The processing-stages assume a sequence and simple relation with clear distinctions, which are not the case in reality (Wickens, 1984, p. 338). The processing-stages assume an explicit representation of an entity (like short-term memory) in the information processing apparatus of the brain (Anderson, 1996). From today's perspective it has to be concluded that human information processing is not built in such a dichotic way.

The sequential information processing approach is a historical burden rather than a suitable approach for cognitive modelling. Virtually any number of stages can be achieved based on the variations in the experiments. The model outlined in the figure should consequently be treated as a description of a structured list of aspects that should be considered if one speaks about cognitive behaviour rather than being a real model of the functionality of the brain. Moreover, this scheme is not a model of human cognition in the engineering sense, because the factor attention, for instance, only influences the processing in an undefined matter by putting it as some kind of 'cloud' into the information processing structure. Even if attention would be a somewhat distributable resource, it would still not be clear from a sequential information-processing model how it is triggered by perception and how it is regenerated. In addition, the distinction between short-term memory (STM) and long-term memory (LTM) is more artificial rather than having a neurological correlate (cf. Neumann, 1992; Strube, 1990; Jones et al., 2003). However, its presence in virtually any textbook of psychology, its simplicity and its applicability in an engineering sense make it difficult to abandon it.

The experiment of Schneider & Shiffrin (1977) eventually falsified the sequential information-processing model. They systematically investigated the habitual aspect of repetitive tasks. People had to perform a search task for a certain number in a set of numbers presented. In one condition, the number to search for was kept constant (consistent mapping condition). In a second condition, the search item varied (varied mapping condition). The outcome of the experiment was that search tasks in the consistent mapping condition are performed much faster than in the varied mapping condition. The results could not fit the paradigm of the additive factor logic and finally falsified the entire theoretical building of the sequential processing model. People start to perform a task in an automated and parallel way during the consistent conditions, which contradicted the prediction of the additive factor logic.

In related studies, Kahneman & Tversky (1979) or De Groot (1966) found that the nature in information processing changes from a more sequential to a more parallel processing mode. Heurisms, efficient search and decision rules are applied to select those aspects from the perceptual side that are relevant for the execution later on. The entire set of perceivable elements is never processed.

The aspect of heurisms has a striking relevance for system safety (Mosneron-Dupin et al., 1997). Four safety critical types of heurisms may be distinguished:

- Availability bias: Human decisions are not based on all information required for a thorough decision but on those available in a given situation.
- Representational bias: Those aspects currently present in memory guide the perception, rather than the objective information given.
- Confirmation bias: Once a hypothesis is built up, one is searching for confirming information, and disconfirming information is neglected. The same holds once a decision on a particular problem solving strategy is taken. One sticks to the problem solving strategy rather than changing it even if the strategy may result in serious problems.
- Central bias: If two mutual exclusive or distinct aspects have to be integrated, one tends to find a compromise solution rather than to choose one of the extremes.

Tunnel vision and fixation in critical situations are known extreme effects of heurisms. The less time available, the more people relate their entire information processing to those situational aspects that are judged as being most relevant for solving the situation. In addition, the memory span reduces in such a situation due to the focused attention (Bubb & Sträter, 2005).

Integration using Phases of Information Processing

One serious problem of considering human information processing sequentially is the fact that the different stages of the model are strongly interrelated and cannot be regarded independently. An approach to overcome this problem and to integrate parallel processing and heurisms with the sequential information processing categories dominating the views in those times is the phase-oriented approach of Rasmussen (1986). The model integrates the aspect of automated human behaviour by introducing shortcuts in the sequential information processing approach. Therefore, the findings on parallel processing and heurisms were integrated into the stage-based approach. The result is the so-called ladder-model, as represented in Figure 2.4.

The nodes in this figure represent the conventional sequential information-processing model. The parallel processing is represented by the shortcuts between the sequential nodes (e.g. from activate to execute in order to represent highly automated cognitive behaviour). Each node has different paths to proceed and therefore decisions are required to decide which path to follow. Heurisms are related to the nodes as those decisions-habits producing the shortcuts in the sequential model. The basis for the ladder-model is the learning-phases that Fitts (1954) proposed for the learning of motor control behaviour.

Depending on where the shortcut takes place, three cognitive levels may be distinguished. The skill-based level is characterized by automated processing of sensory and motor information and needs a high level of practice and experience.

A number of automated ways of behaviour are combined to form a new pattern of behaviour on the rule-based level. Information processing on the knowledge-based level occurs in unpractised or novel situations. The knowledge-based level represents the classical sequential information-processing model.

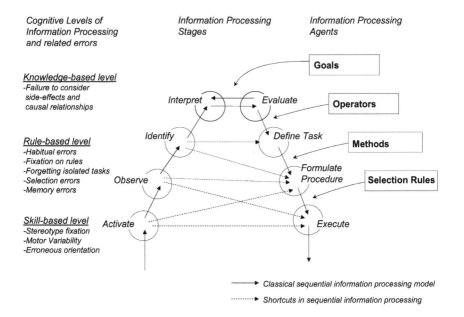

Figure 2.4   The ladder-model (according to Rasmussen, 1986, p. 7)

The ladder-model reflects best practice in the scientific modelling process. It builds upon the classical sequential model but includes new results on automated parallel processing and heurisms. This makes the approach powerful and many other approaches can easily be related to this model as well. The distinction between skill- rule- and knowledge-based behaviour reflects the psychology of learning (e.g. the learn-phase model of Fitts, 1964; Schmidt, 1975). As the figure also shows, the GOMS model (Kieras & Polson, 1985) can be related to the ladder-model. The phase-based approach is therefore able to integrate goal notation methods and object oriented methods. A diagnosis may be described, for instance, by the terms of topographic (top down) and symptomatic (bottom up) search. The reaction time models can be depicted in this approach as well. The shorter the path in the processing stages the faster the reaction time. Skill-based reactions are in the range of milliseconds, rule-based reactions in the range of seconds and knowledge-based behaviour in the range of minutes to hours. This modelling aspect fits the experimental findings that sequential processing takes more time than automated stimulus response reactions.

The Mishandling of the Ladder-Model in Design and Safety

Despite the fact that the ladder-model allows integration of many different findings in cognitive science, it still has the historical ballast of the techno-morphologic modelling age. The shortcuts indicate that no evaluation or formulation of procedures takes place in highly skilled actions, for instance. However, this assumption was already falsified by the experiments of Schneider & Shiffrin (1977). The consistent condition trials did show that people evaluate the stimuli. However, the evaluation or formulation of procedures takes place implicitly or is automated (cf. also Anderson, 1996). The change in cognitive behaviour is therefore rather an implicit automated use of higher-level cognitive processes than a skipping of the processing stage. The ladder-model thus persists in the thinking of sequential information processing stages. This ballast has a tremendous impact on the way the model is used in practice:

- Tasks are equalized with the level of cognitive behaviour (e.g. using procedures means the task is rule-based).
- The assumption that the levels of cognitive behaviour are constant (e.g. independent from a situation).
- The fallacy that skill-based behaviour is safer than knowledge-based behaviour (as assumed in classical human error assessment methods).
- Higher-level stages are ignored as contributors to safety and incidents (e.g. lack of considering goals and intentions for rule-based actions).

The level of cognitive behaviour is equalized with the nature of the task The ladder-model describes cognitive behaviour. However, the levels are treated as given properties of an existing task in engineering. The main first step of the HCR (Human Cognitive Reliability) approach is, for instance, to classify the cognitive level based on the nature of the task and the procedural support given. If personnel has procedures available, this is already classified as a rule-based cognitive level (see Hannaman & Spurgin, 1984). The seduction to allocate the cognitive level the personnel uses to accomplish a task based on externally given information can by no means be concluded from the ladder-model.

Any task is a complex construct (i.e., a hypothetical and abstract term that is related to observable entities or properties but that cannot be defined properly; Dorsch et al., 1994). It may be seen at various levels like the general system-level (e.g. availability of plant), the requirement for a complex decision with major consequences (e.g. shut down a plant) or the operators' perceptual and motor-activities (e.g. alarm checking, switching of a component). At the lowest level, there are very detailed subtasks (e.g. 'move hand to target') reaching the skill-based level. Tasks may also include the awareness about consequences after executing an action (e.g. receiving inquires from management).

Depending on the understanding of what a task is, virtually any level of behaviour may be achieved for any task. While the decision to shut down a plant may involve the entire sequence of knowledge-based processing, the act of switching a component may be skill-based. Because any task may be decomposed into subtasks at any level of detail, any task contains knowledge-based as well as skill-based aspects. Hence, every task may be classified as skill-based, rule-based or knowledge-based, depending on the level of abstraction chosen by the analyst. This holds for all tasks, independent of the working environment (nuclear operators, pilots, air traffic controllers' but also for highly skilled workplaces such as manufacturing lines).

However, this equalization of the cognitive level as an inherent part of the nature of a task continues to exist. In safety assessments, all tasks are usually categorized into skill-based, rule-based or knowledge-based. Knowledge-based tasks are only assumed if there is a very critical situation to manage (Reer et al., 1995). Even highly demanding cognitive tasks of diagnosing a loss of coolant inventory accident (LOCA) event are treated as rule-based actions, because staff members are assumed being trained once a year and having guidance by procedures. The HCR model is still used in some countries for safety assessments, although even the developers do not recommend further use of the model (Spurgin, 2004). Considerable mis-assessment of the human impact on system safety may be the result of such equalization.

In the introduction of new systems, the user it is assumed to handle the system with his or her skills and rules. However, the learning curve for new systems starts at the knowledge-based level and develops into the skill-based level through training and practice. This transition is often neglected in the system design or the training of persons. Because new systems do not change the task, the new mental tasks stemming from the features of the new system are still judged as being skill-based or rule-based, even though they are not.

Summarizing, the levels of information processing depend on training, experience and task frequency and are therefore more a behaviourism description of human information processing than real stages (cf. also Wickens, 1984). Nevertheless, the classical design philosophy is that people can be trained on virtually all systems, as the ladder-model suggests.

The assumption that the levels of cognitive behaviour are constant over time or over situations   The task-oriented interpretation of the levels of cognitive behaviour implies a second assumption in that the levels of cognitive behaviour are stable over time. This assumption is a general one in all safety assessments. However, a task is not performed on a particular level, e.g. the skill-based level. As discussed, the decomposition of a task always leads to skill-, rule- or knowledge-based elements of a task. Rules are combined and knowledge is accessed based on the actual content of memory and based on the experience of the operator in a given situation.

Certain behaviour has to be learned and practised and hence any task goes through all cognitive levels, starting at the knowledge-based level. A classical example is learning to drive a car. While being conscious, cautious and slowly evaluating different options at the beginning of the first driving lesson, the experienced driver performs the same tasks unconsciously and automatically. On the other hand, skills or knowledge may deteriorate due to entropy. Such transition periods and context sensivity of the cognitive levels are usually neglected, although the original reference of the ladder-model describes the relevance of learning for the cognitive levels.

Concluding from the operators' cognition point of view, no single cognitive level may be identified and cognitive skills vary over time. Any simplification of the Rasmussen model trying to use these three levels as task equivalent or stable over time may cause a safety critical ignorance of important error contributions.

The fallacy that skill-based behaviour is safer than knowledge-based behaviour In many design processes regarding safety, skill-based behaviour is predetermined as being safer than knowledge-based behaviour. General statements of this kind imply the design principle that knowledge-based tasks of users have to be avoided (e.g. by using automation) and that training of the tasks to be performed is essential for safety operation. Can such a conclusion be true considering the above? Is this really what needs to be done for error prevention?

Skill-based behaviour can be as unsafe as knowledge-based behaviour. The reason to designate better safety performance to skill-based behaviour is again the equalization of the task with the cognitive level. We assume that the required cognitive behaviour exactly matches the task description foreseen for a certain event. If a system disturbance is happening exactly the way anticipated by emergency procedures, the training and skill-based behaviour is certainly safer than knowledge-based. These types of accidents are so-called 'design-bases' accidents, where potential safety problems (e.g. a loss of a pump in a nuclear plant or a loss of an engine in an aircraft) are identified as potential failures of the system that imply certain mitigation actions from the human side.

However, accidents are always containing conditions that are beyond the design-bases (e.g. Perrow, 1984). Slight deviations of system parameters or additional conditions not foreseen by design may lead to situations where the trained actions and foreseen procedures do not match the situation any more or may even be counter-productive and harm the safety of the system.

Another example is the aircraft accident of a British Midland plane in 1989. A fan blade failure had damaged one engine and the crew mistakenly performed a precautionary engine shutdown on the wrong engine. The damaged engine lost power shortly before landing and the crew was unable to add sufficient thrust to the good engine prior to striking the ground. Forty-seven of the 118 passengers were killed (Airsafe, 2002). Overtraining led to the decision of the pilots to switch off the wrong engine. They were trained to combine a certain acoustic symptom with an engine malfunction. Because a noise, similar to the sound of the trained acoustic symptom, came from the side of the working engine, they diagnosed the wrong engine as failed and switched off the working engine. This resulted in the loss of all engines and a loss of the aircraft.

The safety impact of the cognitive level has to be determined by considering the contextual conditions. The ladder-model (or any other model focusing on modelling the cognitive processes and not considering the context) will not be sufficient to judge the safety impact of human cognition. Recommendations derived from the ladder-model for improving human performance in safety by training of specific tasks are not necessarily adequate. Knowledge-based behaviour may be safer under certain contextual conditions.

Higher-level stages are ignored as contributors to safety and incidents Even if the context is taken into account, the equalization of tasks and the cognitive level in practising safety assessments leads to the dangerous ignorance of the contributions of higher cognitive levels to safety if this model is applied in an engineering sense. It is, for instance, difficult to imagine from the ladder-model that trained personnel will circumvent procedures, because the shortcuts suggest suppressing the influence of goals and intentions of a trained user and treating an experienced user always as information-driven and task-driven.

However, accidents falsify this view. The crew committing the wrong action and circumventing the reactor protection system in the Chernobyl accident was the most effective and highly trained crew, but circumvented essential safety rules in order to accomplish a test.

However, the ladder-model lets these influences appear to be not relevant for skill-based and rule-based behaviour because the interpretation and evaluation phases are shortcuts and therefore there should be no influence of goals and intentions. As a consequence, safety assessments do not sufficiently reflect the influence of goals and attitudes (OECD, 2002). This fallacy is also reflected in many attempts to mitigate intentional human errors after incidents. Violations are far too often tried to be 'trained' or 'taught' away.

If violations occur, they are difficult to be accepted by regulators and event investigators, because qualified persons are not supposed to show this behaviour. Therefore, violations get a negative image as a deliberate act and a purely human-related decision to behave not according to the task. However, again, the context is critical to understanding why people violate procedures. In particular, operational comfort, validity and quality of procedures, as well as unnecessary complexity of the working environment, are contextual factors leading to violations (Shell, 2002). Under poor conditions, like the unnecessary complexity of the working environment, people start to change their goals from simple task accomplishment to task optimization. They generate intentions to overcome the local difficulties and to optimize their working environment. The optimization strategies can become routine in long term.

A typical example is the optimization of permitted walking routes in industrial areas. If permitted walking routes are cumbersome, people start to violate the permitted routes and take shortcuts. Consider a maintenance person in a power plant who has to walk several times between the local settings (to perform his maintenance on a couple of redundant sub-systems) and the control room (to get the work-permission). He will attempt to get all working permissions for all redundancies at once to avoid the walk between the local settings and the control room, even if the systems are required to be treated independently (e.g. in the case of four redundant diesel-generators). The effect of such violations is that safety barriers fail and safety critical events are produced.

The way violations are currently modelled is insufficient. Routine violations, violations on the skill- or rule-based level are understood as deliberate illegal acts, because the input of intentions is assumed to come from the knowledge-based level and to be a result of deliberate decision-making. The role of goals and intentions, which are for instance generated to overcome cumbersome contextual conditions, is insufficiently represented.

On the other hand, deliberate intentional task-performance is only assumed for the rule-based or knowledge-based level. At the skill-based level, only unintentional errors, such as an oversight due to inadequate attention for example, are possible. Consequently, errors of commission, which are located on the skill-based level and are based on intentions, cannot be represented in the ladder-model. On the other hand, such commissions are observed in events where peoples' intentions affected the skill-based level (as for instance in Chernobyl or the mid-air collision or the TMI event where the commission was manifested in a skill-based interaction with the system). Error correction mechanisms can be found only on the rule-based and knowledge-based levels by comparing the actual behaviour with known rules or analogies. Theoretically, the skill-based level does not allow for error correction, although spontaneous recovery of skill-based errors can be observed in events (Sträter, 1997/2000).

Due to these restrictions, the ladder-model could not establish a shift of thinking towards the important aspects of violations. It is still too based on the classical sequential information processing approach. The ladder-model does not explain how memory contents are actually automated and shifted into a higher cognitive level and how the cognitive systems maintain the allocation of attention between all three levels in detail and how they are steered to the 'right direction' (i.e. the resources that are suiting intentions and goals). In other words, if one intends to understand the active cognitive behaviour of humans, one has to understand the process of 'switching between the levels' rather than the distinction of the levels. In terms of attentional research, this means that one has to find out how the allocation of attention takes place (e.g. between focused and divided attention; Wickens, 1984). This is one key aspect for integrating the mystical aspects of human behaviour into cognitive modelling and hence into safety assessments, respectively, into safe system design.

Attempts to Overcome the Limitations of the Ladder-Model

Several attempts were undertaken to overcome the deficiencies of the ladder-model. The ladder-model of Rasmussen was combined, for instance, with the goal notation approach by Freese (in Zapf et al., 1989). He related the ladder-model to situational feedback, memory, and goal selection and therefore reflected errors in goal selection to all cognitive levels. Table 2.1 illustrates this relationship. The table shows a 9 field classification of cognitive errors.

The expansion described in the table gets inappropriate when it comes to active cognitive errors. From the historical perspective of how the cognitive levels are derived from the ladder-model, the relation of the ladder-model to goal notations is a tautology that brings a confusion of concepts rather than a solution to the problem. Goals and intentions are already a substantial part of the knowledge-based level in the ladder-model. Still, intentions have no influence on the skill-based level but only on the rule-based or knowledge-based levels. Errors on the rule-based level are of a skill-based nature (e.g. the skill 'routine violation' as an error in the rule-based level).

Table 2.1   Broadening the model of Rasmussen by adding the goal-related aspects

Cognitive cope Degree of attention Cognitive Load Behaviour Level	*Feedback Processing*	*Memory Information*	*Goal Selection*
Skill-based (for example, daily routine action)	Unconscious slip	Slip regarding problem	Slip regarding problem solution
Rule-based (for example, regular action)	Recognition error violation (routine violation)	Error of omission rule-based error (rule-based mistake)	Error of habit attention error (attention failure)
Knowledge-based (for example, irregular action)	Error of judgement (exceptional violation)	Memory failure	Thinking error (knowledge-based mistake) (deliberate violation)

The Principle of Simplicity of Nature Applied to the Cognitive Levels

The ladder-model has its roots in sequential information processing and the additive factor logic paradigm. The use of phases has enabled us to take a couple of aspects on board, which cannot be explained in the techno-morphologic modelling, in particular the unconscious automated type of behaviour, the learning phases and the different reaction times for skill-based, rule-based and knowledge-based behaviour.

Despite its success, the ladder-model still carries the historical ballast stemming from classical sequential human information processing. The further development of the ladder-model into the Generic Error Modelling System (GEMS; Reason, 1990) shows the difficulties of the techno-morphologic roots of the approach.

Reason (1990) integrated the various psychological approaches that had been employed in the error models until then: the subdivision based on the psychology of memory (e.g. Norman, 1981), the approaches of decision-making psychology (e.g. Rouse & Rouse, 1983) as well as the ladder-model of Rasmussen (1986). Reason thus presents a causal error model considering several psychological approaches on a broad theoretical basis. The main new aspect of the GEMS-model is that it broadens the cognitive levels used by Rasmussen in the direction of decision-making theory.

Figure 2.5 shows the structure of the generic error model (Generic Error Modelling System, GEMS) of Reason. The decision-making aspects are represented in this approach as transition points between the three cognitive levels as defined by Rasmussen (represented as hexagons in the figure). The amount of consciousness and the degree of automation of an action are explained by the levels in the ladder-model (i.e. skill-based behaviour is without, knowledge-based behaviour with consciousness). The new aspect in the GEMS-model is it focuses on the change between the levels. The model exploits the decision theory to describe the transition. A cognitive level is increased as soon as a mismatch is observed between the externally given information and the internal inventory to cope with the information. Different problem solving patterns are then applied at each level, ranging from additional checks to the application of schemes to higher analogies). Herewith the GEMS-model extends the explanatory power of the ladder-model. Violations can be regarded as decision errors.

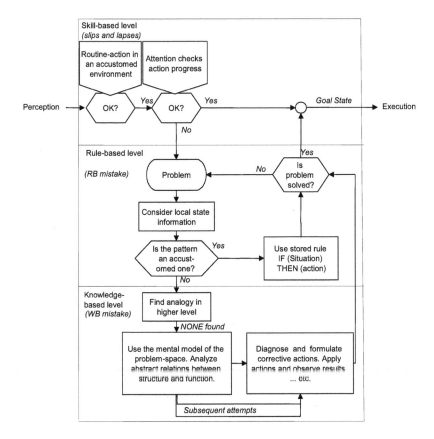

Figure 2.5   The GEMS-model (according to Reason, 1990)

Based on Reason (1990, p. 69), the following basic error types can be distinguished:

- Skill-based level
(a) Perceptual confusion and inattention: Too much information is given to the relevant information for accomplishing the task, leading to stereotype fixation, like focusing attention on familiar information (which also implies inattention on unfamiliar information).
(b) Interference: Confusion of perceptual contents (e.g. visual and auditive information provided).
(c) Repetition and stereotype takeover: Repetitively apply the same skills to different objects as known from compatibility experiments (e.g. Spanner, 1993).

- Rule-based level
(a) Informational overload: Too much information to identify an applicable rule.
(b) Rule strength: The rule is selected based on the frequency of use rather than its valid applicability (strong but wrong rules).
(c) Generality of rules: Application of the same rule to different problems.

- Knowledge-based level
(a) Complexity of the problem: Too many aspects to be taken into account for decision-making and too many causal relationships to be considered.
(b) Biases in reviewing information and goals: Information and goals are not aligned to each other by objective processes but by applying heurisms.
(c) Overconfidence: Once established, problem solutions persist and are repetitively used (e.g. the same problem solution is applied to different types of problems).

Apparently, the mechanisms on the different levels are very similar. Regardless of the cognitive level, common mechanisms for errors can be observed. The items of the (a)-categories reflect very much the bottom-up, information-driven aspect of human errors, the (b)-categories reflect the resolving mechanisms for the alignment of information and goals and for finding a way forward (stereotypes, biases), and the (c)-categories the top-down aspects of generating decisions or behaviour.

The common mechanisms on all three levels within the ladder-model are in contrast to the economic principle of nature. Nature is unlike to build systems with the same error prone mechanisms on three cognitive levels. The principles of nature are:

- Minimize the amount of effort to store and maintain information.
- Minimize the effort necessary to retrieve information.

The principle of simplicity of nature suggests that it is very unlikely that nature develops three similar systems on three different levels. They rather appear to be the result of one central comparison mechanism applied to different levels of abstraction than three different comparison mechanisms. Similar types of cognitive errors happen in the transition from the skill-based to the rule-based level rather than from the rule-based level to the knowledge-based level. For instance, stereotype fixation, rule strength and biases can all be perceived as too strong weights in the neurological correlate (the neural net of the brain). Nature usually finds simpler solutions than such a cumbersome process to maintain information.

Looking at the three cognitive levels from the perspective of economic nature, they seem to be distinguishable by the degree of abstraction and efficiency of using information. The skill-based level deals with chunks, the rule-based level with the combination of chunks and the knowledge-based level with the generating of new chunks. The degree of abstraction is highest on the knowledge-based level and the degree of compilation of information is highest on the skill-based level. A neural architecture capable for representing the three cognitive levels would need a facility to move or transport the information from one level to another while maintaining its content but changing its mode of accessibility (level of abstraction and level of compilation). Nothing is more effortful than moving and compiling information at the same time and hence it is very unlikely that a natural organ like the brain, which is designed to be economic with its energy resources, has separate neurological correlates for the three levels. This also would need a neurological institution responsible for the shifting information and for deciding where to store it. It is easier to leave the information where it is and change the mode of access.

From the error modelling perspective, the assumption of one central mechanism responsible for errors in aligning external information with internal representations and goals would allow a much easier understanding of the cognitive errors encountered by Reason (1990). The influence of goals and intentions on all cognitive levels is easier to model and generic error mechanisms then substitute the arbitrary categorization of cognitive errors into cognitive levels.

An approach of this kind would also ease the understanding of the role of goals and intentions on all cognitive levels and the integration of heurisms into cognitive error modelling. The architecture of such a model will be described in the next chapter. Before doing so, an important further step of modelling cognition in safety should accomplish the short synopsis of this chapter, the role of context for cognition.

The Importance of Context and Cognitive Control Modes for Understanding Human Errors

The term 'Human Error'   If a person commits an error, the causes for the behaviour are not yet known. It could be that the cause for the error lies in deficiencies of the cognitive processing of the person him- or herself (e.g. decay in memory) but it could also be some external cause triggering a certain cognitive process that results in an error. The difference between human behaviour, error and cause for human error is therefore an important distinction to get a model on human error prediction right. It will be shown during this section that the contextual approach is strongly coupled with the cognitive approach.

Hollnagel (1992) distinguished behaviour and cause for the behaviour by introducing the term *Phenotype* for the observable behaviour of humans and *Genotype* for the underlying cause. A similar distinction is made by Bubb (1992) who distinguished between appearance-related and cause-related analysis of human error. The origin for both concepts lies in the classical ergonomics concept of *Load and Cope*. The following definition of human error according to Swain (1992) reflects how a human error is usually defined:

> The term "human error" covers all activities or omissions of a person that result into something either undesirable or that have the possibility of causing something undesirable. ... This definition of the human error is taken in the context of the system, even though the main factors that contribute to an error, for example, can be due to absence of ergonomic design, procedures, training, or a combination of the above. This is why no guilt should be connected with the term "human error".

However, the issue of errors of commission brings forward a third aspect. A human error can actually have two meanings if one investigates an error of commission. The crew of Davis Besse for instance tried to anticipate the reactor coolant injection-pumps in order to save time for further actions. Due to mixing up two switches and a bad plant design, the crew committed the isolation of the steam generators and it was finally close to a core melt situation due to this action (cf. Reer et al., 1999). The action was performed with the best intention to gain time and to smooth the plant operation in this emergency-situation. Therefore, the term error in the above-cited understanding is problematic from the perspective of a human being. An error is an error in the systemic context rather than an error of the human being.

A similar example is the one of the controller who gave wrong instructions in the mid-air collision at Lake Constance (at the Swiss-German border) in 2002. The controller instructed one of the aircraft to descend, while the airborne collision avoidance system on board had already instructed the pilot to climb while descending the other aircraft. The pilot then decided to adhere to the controller's instruction, which resulted in both aircraft descending until they finally collided (BFU, 2004).

In both events, the dynamics of the situation triggered the human decisions and actions. Similar events, where the users try to do the best with the situation and information given but end up in an error of commission, can occur in any industry. Based on such observations, Hollnagel & Amalberti (2001) suggested abandoning the term *Human Error* completely, because a human error is defined by its effect on the system rather than by the human as the cause for the error. Other authors developing methods for assessing errors of commission support this idea, such as LeBot (2004) for instance. Indeed, it would be more correct to define a human error as a faulty state of the system rather than as an incorrect human action in a properly working system. In addition, the discussion of guilt and blame of a human for an error needs to be abandoned, because they are also systemic problems rather than individual ones.

Based on the error of commission issue, a human error should be defined on the level of the system outcome rather than on the individual cognitive level (Sträter, 1997/2000):

> A human error always exists in a working system. It is characterized by an undesired or faulty state of the entire system. It then leads to a situation where the requirements of the system are not met or are met inadequately. The individual is only one part of the working system and interacts together with other portions of the working system. All portions within the working system may be dependent upon each other or may interact in a reciprocal way.

Based on the discussion, two types of contextual factors may be distinguished: The contextual factors of the working environment as employed in the classical human error concept of *Load and Cope* and the dynamic situational factors triggering human decisions and performance in certain situations. These two types are also reflected in the measurements of human performance. Typical measures related to the concept of load are workload measurements like ISA, PUMA or AIM (see for instance Low, 2004). Measurements for the cope-aspect are, for instance, trust or team perception (Kelly et al., 2003; Roessingh & Zon, 2004). The concept of situational awareness is more related to the dynamic situational conditions (Jeannot et al., 2003).

In his method CREAM, Cognitive Reliability and Error Analysis Method, Hollnagel (1998) defines a generic list of contextual factors called common performance conditions. Related to the working environment these comprise: adequacy of organization, acceptable working conditions, adequacy of the Man-Machine interface, and availability of procedures. Related to the dynamic situational conditions these comprise: available time for performing a task, number of conflicting or competing tasks, adequacy of training, and team quality.

Contextual Factors of the Working Environment

The load and cope concept   The classical approach to context is the concept of factors loading the human information processing mechanism. The human then has to cope with the loading aspects. Errors result, if the loading aspects exceed the abilities of the human to cope with them. Consequently, the two types of factors are distinguished in this approach: external factors of the working environment loading the human information processing system and external factors affecting the coping performance of the human being. The factors are usually called Performance Shaping Factors (PSFs) or sometimes Performance Influencing Factors.

The load and cope concept is a further development of the stress/strain concept that was used in ergonomics for the description of the more physical aspects of the working environment (like lighting, noise, radiation). The concept of *Stress and Strain* stems from the techno-morphologic age. The human being represented as a sequential information processor who is rather passively trying to cope with the constraints given in the working environment.

Although the concept was originally not designed to be understood this way, the main point against the stress/strain concept is that it might assume a more or less passive human being only reacting on the stressors he or she is exposed to (i.e., a sequential information processing model of the human being). However, in the course of the developments of cognitive models the concept changed to assuming a more active coping with the stressing aspects of the working conditions. The load/cope-concept reflects the cognitive activities to cope actively with a given situation. *Load* describes the situational characteristics a person is exposed to and *Cope* is related to the active or passive characteristics a person uses to manage the situation. This also includes to actively changing the situation. If the working environment is not flexible enough, this active change may be understood as a violation.

The framework of the Man-Machine System for structuring contextual conditions of the working environment The loading factors of the working environment can be structured well into the framework of the Man-Machine System (MMS). This structure is hidden in some methods and more visible in others. Even in quality or safety management one can find the structure of the MMS, for instance, in the paragraph related to the required resources for achieving quality, respectively safety (e.g. ISO-9000), as follows:

- The availability of internal and external specifications (task requirements) like safety goals, preconditions for operation and specification of results.
- The availability of procedures.
- The presence and use of suitable equipment.
- The availability and use of suitable monitoring tools and measurements.
- The proper realization of checks and measurements.
- Suitable Processes for communication and knowledge transfer.

Such lists can easily be mapped to the structure of the man-machine system as represented in Figure 2.6. The man-machine system describes the following aspects (see also Sträter, 1997/2000; Bubb, 1992). A task is performed by an operator (or group of operators), for instance by receiving an order issue, an order form, or a procedure. After the acquisition of the task and the current system state, the operator initiates his action on the controls of a technical system. The action in turn has consequences in terms of the behaviour of the technical system. The system result depends on the behaviour of the technical system, which, among other properties, is characterized by the dynamics of the process and the degree of automation. The operator gets feedback only indirectly as symptoms of the system behaviour. If the task has been accomplished (successfully or not), the operator usually has to feed back the result to the task assignor (either by filling out a checklist, by signing the work-permission form or by informing his supervisor etc).

The entire man-machine system is influenced by situational factors (like time window) and by environmental factors (e.g. noise, radiation).

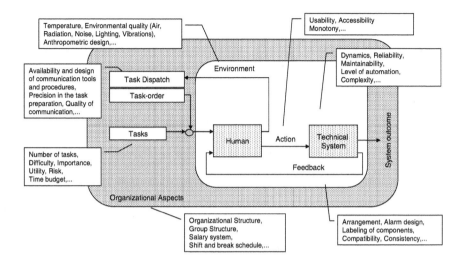

Figure 2.6   Contextual factors of the working environment structured in the Man-Machine System (MMS)

Dynamic Situational Factors

As well as the more static aspects of the working environment, dynamic situational factors play an important role for human decision-making and behaviour in emergencies (Shorrock & Sträter, 2004). Accidents like the above-mentioned mid-air collision at Lake Constance showed that the dynamic aspects of the situation may be much more safety critical than the static contextual conditions. New methods for dealing with human reliability therefore include them systematically into the assessment of human performance. The most prominent developments in this direction are the methods ATHEANA (A Technique for Human Error Analysis; NUREG-1624, 2000) and MERMOS (*Méthode d' Evaluation de la Réalisations des Missions Opérateur pour la Sûreté*; LeBot et al., 1999). Both methods suggest a systematic collection of dynamic situational factors, which can be subdivided into time-related and system-related aspects. The dynamic situational factors can be structured as follows.

Time-related aspects:

Suddenness of onset of a disturbance or system development
- *Sudden, unprepared onset.* Sudden, unprepared onset makes it difficult for the user to get into the situation. A design criterion in nuclear power plants is, for instance, the so-called 30-minutes criterion. This criterion prescribes that the

operators should have at least 30-minutes until they have to act for any design-basis accident scenario.

- *Slow development of system disturbances.* Slow developments of system disturbances are difficult for users to notice. A measure against overlooking events rising from slow developments is the so-called quickened display, i.e. a display showing how the system would develop under the current control parameters in long term (Wickens, 1984).
- *Long period of success or no failure.* A long period of success or no failure may lead to overconfidence in the system and to the false conclusion that a system is stable and error-free. Space and aviation have had to struggle with serious disasters, in particular because management thought the system was safe based on the observation that nothing happened.

Operational phase of a task

- *Reluctance to begin unknown tasks or tasks with drastic effects.* When deciding to perform a task with drastic effects on the technical system (like a reactor shut-down in nuclear power or fuel dump in aviation), the human is usually reluctant to undertake the task (Mosneron-Dupin et al., 1997).
- *Desire of task accomplishment in the late phase of task performance.* Most critical is the late phase of a task system. If problems occur during the phase of task performance where the human has more or less finished the task and just has to do minor things to accomplish the task, any upcoming disturbance is neglected. Experience on this effect can be made throughout any industrial setting. In aviation, this effect is known as the home-base syndrome. Pilots reaching their final destination neglect disturbance more often compared with approaches to intermediate stops or destinations. Doctors tend to forget checking the completeness of equipment in the final phase of surgery. The crew in Chernobyl committed the intervention into the reactor protection system during a rather late phase of the test, shortly before accomplishing the test. Task-driven information processing turns into goal-driven information processing in all these situations.

Involvement of the operator

- *Task takeover.* Humans need time to get themselves into a situation as already stated above for sudden events. This rule also holds for planned task takeover. Kirwan (2002) reported from investigation of human errors in the air-traffic-control environment that the majority of errors were made during the first five minutes after shift takeover.
- *Task release.* Humans also need time to get themselves out of a situation. Effects like the home-base syndrome are the work-related portion of the effect. The same effect acts positively for safety in the after-work phase. During examination of operational experience for instance, failures made during maintenance tasks were spontaneously recovered in the 'after-work phase'. The persons had time to accommodate their experiences to their existing internal world and had released to 'think about the day'. They realized the

error they made and informed the shift on duty. This process was observed as typically lasting about 90 min after the active involvement in the task. After 90 min off work, considerable recoveries have been observed in nuclear power plant events (Sträter & Zander, 1998).

System-related aspects:

Experience with system performance (reliance)
- *Positive experience with the system (overtrust).* If the false alarm rate of a system is low or a system is properly working even in difficult circumstances or circumstances it was not designed for, the user starts to overtrust the system and also own competencies. For instance, overtrust played a role in the Chernobyl accident (the crew was the one with the best performance of all crews) but also in the behaviour of the controller in the mid-air collision at Lake Constance (he relied on the short-term conflict alert, which was put into maintenance that night).
- *Negative experience with the system (undertrust).* If the false alarm rate of a system is high or a system is not properly working even in the circumstances it was designed for, the user starts to undertrust the system and also its competencies. Undertrust may prevent the user from using the system in critical situations or from considering the alarm of a system seriously. During the mid-air collision at Lake Constance, this effect may have prevented the pilot in the Russian airplane from adhering to the TCAS alarm (tactical collision avoidance system) and so deciding to follow the controller's instruction. The TCAS system has a quite high false alarm rate of about 10% and requires decisions of the pilots to judge whether to adhere to the alarm in about 4% of all alarms (Drozdowski, 2003).

Conflicting issues in the situation (task complexity)
- *Conflicting tasks (conflicts).* In the case of additional conflicting tasks in the working environment, the operator has to decide between two equivalent goals. He has to judge the importance of the tasks and to prioritize them. A classical example of a permanent conflict any operator, controller, pilot or doctor is exposed to is the conflict between safety and efficiency or safety and economy in general. Another issue might be inappropriate waiting time. Mehl (1995) investigated this effect in the maritime environment and found that captains start to violate rules to yield other ships, if the waiting time exceeds a certain subjective tolerance band.
- *Dominant problem in a multiple problem situation (masking).* One dominant disturbance may mislead the users in noticing other relevant disturbances. If a steam-generator tube rupture has to be covered by operators in a nuclear power plant, other minor additional problems may be overlooked. Such minor disturbances may become critical in later stages and also need recovery. The effect was systematically investigated in GIHRE (2004) for a range of industrial settings. Note that the simple exceeding of the number of tasks

(overload) would be covered already in the classical load/cope concept as described above.

- *Operational comfort.* If a task contains more effort than needed to accomplish it, the operation becomes unnecessary complicated and operational comfort diminishes. If the normal operational system is not available and people have to use the back-up system, it is highly desirable to design the back-up in exactly the same way as the operational system. A reduced set of functionality may seduce the user to return to the original system even though this failed. In designing work-processes in safety, redundancies are to be checked independently. However, this process means that the maintenance-staff has to ask for four work-permissions to accomplish the required maintenance works for a system with four redundancies. Work-permissions in nuclear facilities are usually issued in the control room and are represented by independent locks. In order to perform the work on the local system this means that a maintenance person has to go back and forth four times between the control room and local system to pick up the required keys and work-permissions for each redundancy. The probable practice is that staff asks for all four work-permissions and keys in order to save the walking time (GRS, 1998a). Similar events can be observed in occupational safety. People violate permitted routes in order to save time to accomplish their work. The shortcuts lead to collisions or accidents, which cannot be covered by training or sanctions. Cumbersome work-processes hence lead to safety violations for the sake of work efficiency (Shell, 2002). They represent about 50% of all costs for health and accident insurance (e.g. Siemens, 2002).

Ambiguity of information in the working environment (uncertainty)
- *Lack of transparency of the system.* Lack of transparency of the system enforces people to make assumptions about the causes for a failure. They consequently build hypotheses that are based on incomplete information about the system, which do not necessarily fit reality. Strohschneider (1997) investigated the effect of incomplete information in decision-making and found that people make decisions on a best guess basis, even if they know the information is not sufficient, that the decisions are usually rather short-term related, and that counter-signals on the validity of the decision may exist.
- *Dependencies in the system.* Dependencies in the system make the prediction of the success of an action difficult. Dependencies are known to cause latent errors, such as making a complete redundancy unavailable by mistake due to dependencies in the subsystem (cf. Sträter, 1997/2000).
- *Ambivalent symptoms.* Symptoms that can be interpreted in different ways will be interpreted wrongly in some instances. This factor is always true if incomplete or ambivalent statements exist in the system, like equivocation of equipment (e.g. plugs are not designed in a way that they can be in inserted without the risk of confusion such as twisted plugs). A doctor may diagnose a simple cold from the symptom 'fever', although it might be a serious influenza.

Misleading information processing (priming)

- *Misleading sequence of alarms or signals.* A misleading sequence of alarms or signals may put the operator, pilot or doctor on the 'wrong track'. If the alarm for a steam generator tube rupture is coming first and a second more important alarm later in the sequence of events (like an additional alarm on radioactivity in the cooling system) the second alarm will be more difficult to handle. The more the operator is actively involved in problem solving the first problem, the less he is able to notice even the second alarm (Fukuda et al., 2003). People start to work on a certain problem-solving path, and any additional information appears to require an additional cognitive effort, which is not undertaken.

- *Misleading decisions due to latent errors in the system.* Latent, not recognizable errors in the system may lead to false assumptions on possible problem solving strategies. An operator decides based on the specified availability of systems. However, the specified availability may not be guaranteed if there is a latent error in the system. Consequently, the operator may initiate an undesired state by performing usual actions or may fail to undertake a successful recovery action. Examples of such traps are numerous. A cable of an emergency pump is designed too thin and is overloaded in long operation. The operator would probably not think about such a construction error of this kind in an emergency and could cause a cable fire in addition to the emergency.

All dynamic situational aspects seem to be related to the misalignment or de-coupling of the human with the task-environment. They all are common in that they cannot be avoided by a classical re-design of the system (like a redesign of the ergonomics, improvement of training or procedures). They rather reflect the inherent problems of any system that exists as from its conceptualization. Such inherent problems cannot be explained in the current models for cognitive information processing that are used for safety assessment. It is obvious that the design deficiencies lead to a circumstance where the task behaviour of the human is not very well coupled with the other aspects of the working environment anymore. The users' performance deviates from the task, from the information provided by the system, from a communication partner, from a piece of hardware, or software. The misalignments have in common that they break designed safety barriers in the system.

Although the dynamic situational factors are rarely represented in safety assessments of technical systems, accidents and incidents have shown for some decades that these aspects are substantial for understanding the actions of the personnel. One reason is that the implemented safety assessments still rely on the sequential information-processing model and the related load cope concept. A shift in the paradigm of safety assessment towards considering the active role of human information processing is needed in order to realize the implementation of the methodological developments in safety assessments. The shift in the paradigm also includes a shift away from the classical approach of safety assessments using fault and event-trees as typically used in PSAs (Probabilistic Safety Assessments). The classical approach represents human performance by failure events in the fault-tree. A fault-tree contains failures to perform the required human behaviour (based on the procedures given) but not the active part of human behaviour. In contrast, human performance problems identified in real operational events often involve operators performing actions that are not required for accident response. These, in fact, worsen the plant's condition (i.e., errors of commission) as shown in some serious accidents, such as those at Chernobyl and Three Mile Island (TMI-2). In the TMI-2 event, operators inappropriately terminated high-pressure injection, resulting in reactor core under-cooling and eventual fuel damage. The mapping of the events to classical safety assessments frequently leaves the impression that the operator's actions were illogical and incredible. Consequently, the lessons learned from such events often are discounted with the safety assessment model. As stated in NRC (Nuclear Regulatory Commission; NUREG-1624, 2000), after the TMI-2 event, numerous modifications and backfits were implemented by all NPPs (Nuclear Power Plants) in the United States, including symptom-based procedures, new training, and new hardware. However, after these modifications and backfits, these types of problems continue to happen. The problems are a result of errors of commission involving the intentional operator bypass of engineered safety features. In 1995, NRC's Office of Analysis and Evaluation of Operation Data (AEOD) published a report entitled 'Operating Events with Inappropriate Bypass or Defeat of Engineered Safety Features' that identified 14 events over the previous 41 months in which an engineered safety feature was inappropriately bypassed. This discrepancy is a result of an improper cognitive model used in the safety assessment and safety engineering.

The observations of the NRC strive for better safety assessment regarding the active role of human behaviour in safety-engineered systems. The active human behaviour has the potential to overrule virtually any engineered safety system. In Chernobyl, even the most important safety net, the reactor protection system, was overruled because the crew was convinced of their ability to control the situation successfully. In the mid-air collision, the controller and the pilot were convinced they were doing the right thing to secure the situation.

The importance of the dynamic situational context is consequently perceived in virtually all newer methods on modelling cognitive aspects for safety (OECD, 2001). As outlined in Part II and III of this book, an integrated cognitive model will serve as a basis for better inclusion of human behaviour in safety assessments and will build the rationale for a human-related safety assessment methodology.

The CREAM method of Hollnagel (1998) made a considerable step forward in tackling the cognitive modelling of the dynamic situational context. CREAM introduced the contextual control model (COCOM) which includes a set of cognitive control modes to describe how human information processing deals with the dynamic situational context. The approach is a shift in the paradigm because it allows for iterations of information processing stages and perceives the cognitive processing as being coupled with the contextual conditions. Hollnagel distinguishes the following control modes: scrambled (occasional control), opportunistic (information-driven control), tactical (planned control) and strategic control (goal or intention-driven control).

Long-Term Effects of Inappropriate Context

Permanently ill-designed systems have considerable long-term effects on safety. They lead to a manifestation of attitudes towards the tasks, systems, communication partners, the management or regulation. Trust is a classic example. Problems of motivation or job-satisfaction in an ill-designed manufacturing plant are known problems in occupational safety. Routine violations or lack of safety culture may be the result.

All these are the long-term effects of dynamic situational conditions, if the conditions are not changed, or properly managed in case they cannot be changed. Psychology coined the term *Learned Helplessness* for describing the manifestation of inappropriate working conditions in attitudes towards the work in general. Learned helplessness is difficult to reverse. Once mistrust into a system or software happens, it is difficult to get the staff to trust the system again (Kelly et al., 2003; Bonini, 2004). It may take several years until the software can be reissued to the staff.

The contextual factor monotony may serve as an example of how such learned helplessness develops. In monotonous environments, vigilance decreases and the inertia to react properly to changes of plant states increases. In order to compensate for this development, users may start to search for change and flexibility. People may start compensating for monotony by changing their working processes, by taking over automated functions to make work more interesting or by creating new challenges. In case this type of job enrichment or variation is not possible, people tend to develop a 'don't care feeling' about the situation.

Long-term manifestations of the improper contextual conditions in attitudes are very difficult to overcome or to change by classical design. Measures like job-rotation or participation of staff in improving the plant or construction lines were proven efficient in the longer term (Schuler, 1995).

In any case, a cognitive model capable of representing the long-term effect of contextual conditions on safety needs to address how attitudes develop and how they contribute to the cognitive information processes.

The Combination and Interdependence of Contextual Factors

Contextual factors are not independent from each other. For instance, even under high workload conditions, humans are able to perform error-free and efficiently (Fukuda et al., 2003). The simple statement 'high workload means high potential for errors' cannot hold. The same is true for the stress concept and others. Psychologists name the variables to take into account in addition to the main relationship *Intervening Variables*. Overall, a multidimensional problem, i.e., a problem of many intervening variables in the relationship of errors and contextual factors is the result. Dependencies between contextual factors reflect two aspects of cognitive processing (Sträter, 1997b).

First, they reflect the interference in the cognitive system. As discussed regarding the experiments of dichotic hearing, words caused auditive and visual interfere so that words from different channels were mixed up.

Cognition and Safety

Second, they reflect specific experiences the human beings made in their working environment. If people know that a system is not working properly in a certain situation, they will avoid using it and, instead of this, they will make decisions based on their experiences. Given the contextual factors and the tasks to be done, they will decide about the utility of certain technical, human or managerial constraints they are faced with.

The problem of the interrelations of contextual factors was demonstrated in Sträter (1997/2000) for the case of causal factors in event reporting. It could be shown that the neglect of dependencies leads to completely misleading conclusions. Figure 2.7 shows how the different causal factors are interrelated. The closer data points the higher is the dependency between the factors. The factors represented in the figure may be summarized as follows (the number of observations is mentioned in brackets; see Sträter, 1997/2000 for the detailed discussion of the interrelations and interdependencies of the factors):

Mission/Task related:
- Task preparation (13): lack in planning, organization or preparation of task.
- Task precision (7): lack of precision of a task.
- Task complexity (6): task was too complex in the given situation.
- Time pressure (6): time pressure caused trouble in performing the task.
- Task simplicity (3): task-deviations in a given situation were not considered.

Communication/Procedures related:
- Completeness of procedures (24): information in procedures missing.
- Presence of procedures (7): procedures do not exist for the given situation.
- Design of procedures (4): procedures are ergonomically badly designed.
- Precision of procedures (3): procedures lack preconditions for action.
- Content of procedures (1): procedures contain wrong information.

Person related:
- Goal reduction (11): task and goal to be accomplished was simplified.
- Information (8): person ignored information provided by the system.
- Processing (7): fixating on a task led to the error.

Cognitive processing related:
- Usability of control (12): control-systems are unpractical to handle for operators.
- Positioning (7): control-systems cannot be placed in a certain position.
- Quality assurance (7): quality assurance was not used.
- Equivocation of equipment (6): equipment can easily be mixed up.
- Usability of equipment (4): equipment cannot be handled by operators.
- Monotony (4): action is monotonous, typical vigilance effects occur.

Feedback related:
- Arrangement of equipment (6): equipment badly arranged.
- Marking (6): semantic meaning of system states not marked on the display.
- Labelling (5): display is not labelled or badly labelled.
- Display precision (4): display does not show process parameter precisely.

- Reliability (2): display is unreliable and hence operators' belief in it is low.
- Display range (1): display is not capable of displaying a process parameter.

System related:

- Construction (8): a system (like a diesel, a conduit) is badly constructed so that it is obstructive for operators.
- Coupled equipment (6): electrical equipment that is supposed to be redundant impacted on each other or is coupled.
- Technical layout (4): layout of system is beyond required specification.
- Redundancy (4): several redundancies are affected (common causes).
- External event (1): an unforeseen external event caused technical failure.

The figure shows for instance that the adherence to procedures (respectively violation of procedures; as expressed with the dot named 'goal reduction') is strongly related to the design of the task organization and the ergonomics of the working environment (e.g. 'completeness' of procedures or 'usability of control'). Consequently, the problem of violations can immediately be seen as far more than an attitude of the personnel as sometimes done in the context of safety culture.

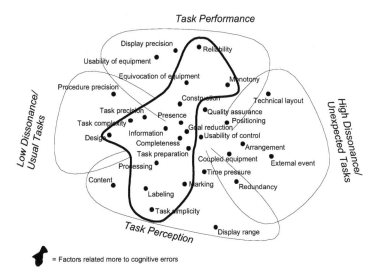

Figure 2.7   Dependencies of different causal factors in incidents (Sträter, 1997/2000)

The way humans build up their knowledge about how to deal with a certain system in a certain context is experience-based. Reason (1990) described numerous examples of how errors occur after re-arrangement of the usual way certain equipment is arranged, because people relate their behaviour to the experiences made. The rearranged location of the equipment has to be re-learned by the staff. Consequently, a cognitive model needs to explain how the dependencies between required or expected behaviour and contextual factors develop. Essential for the explanation of the role of contextual factors and, in particular, of the dynamic situational factors is that humans build up knowledge from their experience and base cognition on an event-based inference mechanism (Shastri & Ajjanagadde, 1990). This event-based inference mechanism needs to represent the multi-dimensional nature of cognitive processing in order to reflect the relationship of cognitive processing and the role of context. It also needs to reflect the different modalities and the role of memory. This requirement leads to the maxim that an integrated cognitive model needs to be event-based.

Part II

Integration

Chapter 3

The Cognitive Processing Loop

Modelling always shows our limited possibilities of expressing all aspects within a concise picture. The techno-morphologic approach discussed in the previous chapter shows how much technological concepts were used to build psychological models. The memory metaphor and the approach of processing stages is the most prominent example for this techno-morphologic thinking. Such approaches comply with some kind of helplessness in finding a psychologically sound approach for modelling cognition. An integrated model needs to be based on more than take over of technological concepts. On the other hand, approaches that are more psychological have been developed. Examples are the SOAR approaches, ranging from goal state models and notations like the Cognitive Complexity Theory (CCT; Card et al., 1983), the Adaptive Control of Thought (ACT – Anderson, 1983) to the PSI model (Dörner, 1999). Another example, coming from a different angle, is the modelling of physiological aspects based on EEG-Electroencephalogram research (overview in Sommer et al., 1998). These methods have additional aspects to offer that complement the existing approaches to safety. However, they did not find a path into error modelling and safety assessments because they were too far away from the engineering needs of safety assessment. The same holds for the detailed aspects of the work of Rasmussen (1986).

Often, the detailed discussions that led to a certain cognitive model are not considered (and cannot be, due to resource restrictions) in practice. Instead of the entire model, often the first-order categorization of a model achieves implementation only. It is a natural process that only a few major characteristics are transferred from extensive scientific investigations into practice. The concern of this chapter is how to find a first-order structure meeting the aspects to be covered in modelling cognition for safety.

Characteristic of an Integrated Modelling of Information Processing for Safety

General Reflection of the Genesis

As discussed in the previous chapter, the current approaches on cognitive modelling for safety still persist in the additive factor logic and the inherent limitations of this approach (cf. Anderson, 1996). Even more advanced models like the GEMS-model (Generic Error Modelling System) of Reason (1990) did not overcome the historical ballast of the processing-stages. One can see the use of processing-stages in virtually any structuring of human-related issues. They reoccur in models for recovery from system failures (Shorrock & Sträter, 2004) as well as the structure of causal analysis sheets, the fault-tree/event-tree approach of safety assessments, questionnaires or dynamic models of human performance etc.

Information-processing stages were derived from so-called degraded-stimulus experiments. In those experiments, people were forced to recognize digits with reduced readability in a certain amount of time (Wickens, 1984). The cognitive processing stage that was assumed by the experimenter was approved in the experiments and, if statistically significant, assumed to be a stage of processing that is influenced by the degree of degradation. This logic was named *Inquisitory Logic* because the resulting structure of the cognitive model is a reflection of the experimenters' expectations rather than a model of the piece of interest.

Those models still reflect the classical sequential model somehow or, in other words, rely on the primacy of the perceptual system (Messing, 1999). They will be called *Techno-morphologic Models* hereafter.

The previous chapter discussed that no sequence of processing-stages can be assumed if one looks at the errors happening in accidents. Information processing is apparently not a linear relationship or sequence between input and output and not explainable by models based on the techno-morphologic thinking. Searching and looking for relevant information in a disturbance situation is strongly related to active, intention-oriented processes in the brain. In order to overcome the historical ballast of sequential processing modelling, the lessons discussed may be summarized as follows.

- The processing-stages are interrelated and are not distinct from each other. Cognitive behaviour is dependent on various mental processes and is not an independent set of distinct processing-stages. Response planning has also an impact on situation detection and assessment. Humans steer attention to what they know (expect) or what they are able to do. Well-known investigations concerning this problem were performed some time ago and led to the concept of cognitive interference. Broadbent (1958), for instance, found high interference between memory and recognition in dual task paradigms and tried to explain this with his famous filter-model.

- The processing-stages are not always explicit. The most serious problem as discussed in the context of the ladder-model is that the stages are not always explicit. They are implicitly involved in the cognitive processing rather than shortcutting the processing stage. This behaviour is well investigated in the experiments of Schneider & Shiffrin (1977) or Anderson (1996). Within these experiments, it was found that people automate their behaviour without changing the information-processing stages that were used for cognition, but accelerate the information processing. In accordance with the compilation of a computer-program, Anderson described this phenomenon as compilation of information. This phenomenon of compilation of cognitive behaviour is currently poorly considered. Instead, the concept of skill-based behaviour, for instance, is only realized as shortcuts in sequentially organized processing stages. The idea of shortcuts does not allow that all information processing stages are involved in cognition and hence higher-level contributions to risk are neglected.

- The processing-stages and ways of processing are not able to describe known mechanisms of cognitive behaviour completely and sufficiently. Focusing on processing stages, as the major parameter to structure a model, leads to the issue that other aspects of cognitive modelling are only included 'second class'. The ladder-model, for instance, is able to describe search in diagnosis by the terms topographic vs. symptomatic search, but on the second layer only. The same holds for psychological findings concerning memory (e.g. limits of short-term memory, influence of long-term memory on search, perception and recognition). Finally, the influences of the endocrine system, which is responsible for more general influences on cognition (e.g.; fatigue due to attentional resources, tunnel vision due to stress or fear), come on the third layer of explanation (cf. Rasmussen, 1986). Herewith important aspects of cognition are not immediately visible on the first layer of the models and are easily lost in practical applications. Only the high-level structure seems to find its way into application. The example of how the GEMS-model is practised is a good example of how little a model is implemented thoroughly in safety practice. The underlying rationale of the top-level structure of the model remains usually unconsidered.

Conclusion from the Genesis of Modelling Cognition

We are currently still very much bonded in the history of techno-morphologic information processing modelling. This view on cognition steers the understanding of human error with a tremendous impact on the understanding of how violations emerge and how they can be prevented. In addition, the search for causal factors in incidents is very much a result of this genesis as well as the thinking on design measures to overcome human error. The historical view is not only manifested in simple measures, like precautions on the operational level (training, procedures), it is also steering our way of thinking in the regulatory requirements or safety assessments (e.g., the human is modelled as an information processor in fault-tree/event-tree based safety assessments). Finally, the history of the techno-morphologic information processing gives a rather reactive view on human cognitive processing. As a consequence, safety assessments and design suggestions regarding human error are more reactive as well, because they follow from the cognitive model. The reason why the obsolete approach is still in practice is the considerable sluggishness of getting research into implementation. Usually one has to expect about 10 years until research results are implemented in applied industrial settings.

Another aspect to highlight is the suppression and neglect of important aspects of human nature in the current cognitive approach for safety or design. It is often criticized that the positive aspects of human performance are suppressed by design recommendations coming out of safety assessments as well as emotional aspects. The issue of error of commission cannot be tackled either. The historical ballast makes the integration of the active parts of human cognitive processing into the general safety-approach cumbersome, less visible or even impossible.

A considerable step forward was undertaken by Reason in distinguishing unintentional error from intentional error (Reason, 1990). However, this important development could not successfully find a path into safety assessments and design because it was 'clapped on' the paradigm of the sequential information processing and did not overcome its limitations.

An additional step forward is the concept of cognitive control modes introduced by Hollnagel (1998) and emphasizing the coupling of cognitive processes with contextual factors. The idea of coupling is an essential step forward in overcoming the reactive techno-morphologic information processing approach. The coupling allows representing the most essential human cognitive ability, which is the reflection of one's own thoughts in order to generate new ideas (e.g. Michel, 1989).

The discussion of the mystical active nature of human information processing showed further that the active human behaviour is strongly connected to the way experiences are turned into memory and how memory is used to formulate goals in a given task-environment. The lattices approach, a connection matrix for human signal response relations, of Moray (1990), and the perception action model of Bubb (1992) are outlining how to integrate task, memory, goal and experiences. Jones et al. (2003) formulated the short-term as 'low probability transitions' of contents in the long-term memory and concluded that the difference in transition probability is reflecting the difference between STM and LTM better than two storage devices as it was concluded from the computer-metaphor.

In order to find safer design and better assessment of the human characteristics, assessing of human performance based on human characteristics is needed rather than on oversimplified models suiting superficial engineering needs. Essential aspects to learn from the genesis are:

- Recognition of the valid aspects of techno-morphologic information processing aspects.
- The inclusion of intentions, goals and attitudes.
- The inclusion of heurisms and utility.
- The harmonized and integrated view of short-term and long-term memory and reaction time.
- The impact of dynamics of the control loop on higher cognitive functions.
- Simplification of cognitive levels by assuming central mechanisms and levels of abstraction.
- The continuous nature of levels of cognitive processing.
- The coupling of information processing and context.
- The practicability of cognition in safety assessments.
- The requirement to overcome guilt-related error classifications.
- The requirement of building an explicit scientific process of human error modelling.

Recognition of valid aspects of the techno-morphologic information processing
Although the techno-morphologic information processing has the limitations
mentioned above, the general structure of having some input, a cognitive
processing and some output are to be maintained from the techno-morphologic
approach, because they reflect the physiological distinctions of sensory system,
brain and executing organs. Input and output have to be understood as interfaces of
cognition to the external world.

The inclusion of intentions, goals and attitudes The influence of intentions, goals
and attitudes on information processing was striking from the accidents discussed.
Intentions, goals and attitudes have the potential to steer perception, decision-
making and action and herewith to break the designed safety barriers of a system.
They relate to all processing stages and are not restricted to those of the
knowledge-based level. An integrated model therefore has to allow for the
influence of intentions, goals and attitudes on perception, for instance.

The inclusion of heurisms and utility The events discussed reveal the importance
of heurisms and utility in the decision-making and problem-solving act. In the
ladder-model, they are only represented by the distinction of the three cognitive
levels. This forces us to define heurisms or utility differently for each level. This
results in an inherent conceptual contradiction in that heurisms (as being of skill-
based nature by definition) have to be assumed as essential limitations of the
higher rule- or knowledge-based level. The inclusion of heurisms therefore breaks
the inherent logic of the ladder-model. In the same way, utilities are assumed to
come into play only on the knowledge-based level but affect the skill-based level
as well. The role of the more unconsciously applied heurisms in cognitive
behaviour needs to be better aligned with the more consciously applied utilities in
order to understand the bounded rationality of the problem-solving act (Simon,
1955).

*The harmonized and integrated view of short- and long-term memory with reaction
time* The timing of reaction (as being modelled via shortcuts) would not allow for
spontaneous conscious decisions in skilled performance as represented in the
accidents discussed. For instance, the action to anticipate the emergency cooling in
the Davis Besse event and an intervention of a controller into the airborne collision
avoidance are difficult to represent in the ladder-model. So far, slips like the ones
discussed are considered as unconscious and random. However, they contain issues
of memory, goals and experiences. A harmonized view on how tasks, memory,
goals and experiences are interrelated and how these are managed in terms of
timing during emergency situations is essential to understand the nature of slips
better. This view includes discussing the distinction of consciousness and
unconsciousness and its relation to short-term memory, which appears to be a
memory span in a given time rather than a storage device with time-independent
characteristics (Katzenberger, 1967; Card et al., 1983).

The impact of dynamics of the control loop on higher cognitive functions Higher cognitive functions like decision-making or calculating are dependent on the dynamics of the tasks (Card et al., 1983). The higher the dynamics the lower the memory span. This could be confirmed by a number of studies on car-drive capabilities to manage complex menu-structures of navigation systems while driving (Rassl, 2004). This interdependence of time and memory span is critical in all dynamic task-environments. Air traffic management represents such an environment where increasingly complex and automated interfaces are introduced with the hope of increasing the capacity of the controller per sector. However, it is assumed that the introduction of the new systems has only positive effects on handling the number of aircrafts. The impact of the additional interactions with automated systems on the memory span (and hence on the number of aircrafts to manage at a given time) is neglected. Finally, this reduction of the memory span has an impact on the decision-making capabilities and response times. Again, such impact is not represented in the ladder-model. The shortcutting of higher cognitive functions explains the faster reactions of skill-based behaviour (in the range of seconds) compared with rule-based behaviour (in the range of minutes to hours) but not the impact of skill-based dynamic control on decision-making capabilities.

Simplification by assuming central mechanisms and levels of abstraction Within the levels of cognitive behaviour, one can see three times the same general cognitive properties just on different levels of abstraction (biases in information reception, in aligning information and intentions, and in generating decisions or behaviour). This structuring is sub-optimal from the scientific criteria of simplicity of nature. The structure of the cognitive model seems to be flatter. It is very unlikely that nature builds a system three times with similar deficiencies just on different levels of abstraction. Biases seem to be more a property of a central mechanism that facilitates the access to the information in decision situations. The striking similarity of error mechanisms on the three levels is strongly suggesting one central mechanism that holds for all attentional allocation of perceptual tasks, rule strengths and biases. This simplification makes it much easier to understand how errors occur as if one assumes the cognitive levels are different levels of abstraction. This assumption also eases the integration of interference effects and modality-effects of memory throughout the spectrum of cognitive behaviour (skill-, rule- or knowledge-based).

The continuous nature of levels of cognitive processing The cognitive levels used in the ladder-model are not as distinct as they appear but rather represent a continuum (e.g. Wickens, 1984). Cognitive levels depend on the degree of practice achieved by an individual and are thus specific to the person and not specific to task characteristics or the situation the person is in. It is therefore not too surprising that Moieni et al (1994) were unable to prove the cognitive levels empirically Instead, they found subdivisions along the system characteristics and dynamics.

However, current modelling of cognition for safety by fault-trees and event-trees requests modelling cognition in a discrete way, which contradicts the more continuous nature of cognition. This restriction of safety assessment methods has to be recognized but a cognitive model should also overcome this limitation. In order to get a usable approach for dealing with cognition in safety, the model should be capable to respect the current discrete modelling. However, cognitive modelling should also enable more continuous modelling because dynamic risk modelling is increasingly introduced into safety practice (Mosleh & Chang, 2004).

The coupling of information processing and context The issues discussed as a dynamic situational context are not covered in current modelling of cognition in safety. Conversely, they were of course investigated or modelled in other scientific work. This work did not find its way into practice. The practical use of the models from Rasmussen or Reason shows the cause for this lack of implementation. Although the ladder-model and the GEMS-model are well aligned to safety needs, only the high-level structure of the models remains in the practitioners' minds. A model intending to cover the dynamic situational context consequently needs to represent these aspects already on the top level of the model.

The reflexive nature of human information processing and active commissioning of errors is only understandable from the way context and cognition is coupled. Cognition is an experience-based cognitive system. It combines experience and characteristics of human information processing and represents the dependence of behaviour on context, from which behaviour is triggered. Contextual control modes were introduced by Hollnagel as a general approach for describing the resolving mechanisms of information- and intentional- or goal-driven processing.

The practicability in safety assessments The judgement whether a human action is safety critical is based on the model of cognitive behaviour we use. The techno-morphologic way of thinking is still the kernel of the way we currently deal with human impacts in safety and the aspects mentioned above are usually less reflected in error modelling. However, these aspects have to be contained in a cognitive model in order to face better system design and assessment in the future. The conclusions from the current use of modelling cognitive information processing in safety assessments and design are dangerous from the safety perspective. Theis (2002), for instance, encountered that not dealing with the dynamic situational factor 'conflicting tasks' leads to a mis-assessment of the error probability by a factor of about 200.

Safety assessments are remote settings of cognitive modelling. It is neither feasible nor possible to predict the cognitive level an operator might have during an accident sequence (Sträter & Bubb, 2003). Intensive queries of the experience levels of the potential operators and the potential contextual conditions would be required and the perceptual units of the operators have to be identified individually (Sträter, 1991). Safety assessments, however, deal with an expectation for a set of persons rather than individual performance.

The requirement to overcome guilt-related error classifications The cognitive model used to describe and analyse events or to assess the safety impact of humans on safety is also a basis for the judgement about the responsibility of the human for an event or an accident. Any model is therefore also an approach to explain human behaviour. In the sequential information processing approach for instance, the errors assigned are related to the processing-stages as the underlying psychological model of cognition (e.g. failure in perception, wrong identification, failure to note, wrong interpretation, incorrect understanding, wrong evaluation, incorrect diagnosis, wrong execution, incorrect adjustment, interpretation, remembering, or execution). In more advanced models, they can be understood as combinations of these generic cognitive stages (e.g. failure in reading, thinking, calculating, etc). Those assignments are more or less settling the potential causes and therefore the potential responsibility and guilt.

Often, the stages of human information processing are used for cognitive error classification, because these are simple and easy to understand. A blame-free dealing with incidents, as often requested, as basic feature for good safety culture, requires a shift to a cognitive model that serves as a basis for better understanding and assessment of the human role in incidents.

It is important to overcome the current safety approach regarding cognitive issues in order to found a blame-free culture. The techno-morphologic information processing models inherently contain the idea of blaming the cognitive system of the one who made the error. If a blame-free culture and a guilt-free analysis and assessments of cognitive errors are to take place, a shift in the metaphor needs to be made away from models based more or less on the additive factor logic (Hollnagel & Amalberti, 2001). If we persist in modelling cognitive errors by assigning error modes like failure to note, incorrect interpretation, misunderstanding, lack of use of knowledge, bad diagnosis or incorrect adjustment we will never be able to overcome thinking in simple causes like lack of training or bad procedures. Finally, we will never be able overcome the limitations of current safety assessments regarding cognitive errors. As a consequence, accidents and disasters will not be analysed sufficiently regarding the cognitive characteristics of humans and finally countermeasures will never be effective enough to prevent human-induced accidents.

The requirement of being scientific   In conclusion, the current status of how cognitive issues are currently developed and represented in safety assessment as well as in design for safety is hardly to be based on a scientific background. Rehearsing Popper's example of the black swan in the set of white swans, the accidents happening and cognitive effects observed (black swans) contradicts the models used for assessment and prevention (white swans). The models used for safety assessment need to be revised continuously and systematically based on such experiences in order to complete the current safety approaches. The discussions in this chapter have made clear that this also includes research and regulatory philosophies.

The General Nature of Cognitive Modelling

There is a relationship between the perception of coincidence and the model of cognition we have in our mind. Those aspects not represented in a model are usually perceived as random events and coincidence. We do recognize only after some time and after a couple of more or less critical events happened that the model in our mind represents the observations incompletely.

Iterative Nature of any Scientific Loop of Progress

In order to understand why cognitive modelling does not suit system design and safety, it is worth looking at the construction process of those models that found their way into system safety (in particular into safety management, assessment approaches and design modelling approaches).

The basis for all cognitive modelling is observation of behaviour of persons in certain environments. Observations may take place in controlled environments (experiments) or uncontrolled (field investigations, events, accidents). However, it is not the case that observations are purely data-driven. They already require some kind of cognitive model for preparation of the experiment or for structuring the field-observations, even if this cognitive model is a basic or rough one. Data and models have a non-dividable liaison and iteratively develop each other further (Figure 3.1). A special type of the data-driven approach is the direct observation of brain structures to derive hypotheses about the way how the brain works.

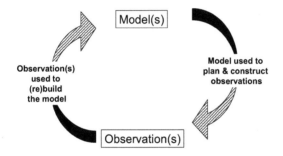

Figure 3.1   The iterative liaison of data and models

In general, there are two possibilities to make predictions about human behaviour. The first approach is the data-driven approach to take the observable information and to derive hypotheses about the underlying information processes. The second approach is the model-driven approach to develop a hypothesis or cognitive model about how human information processing works and to prove this in an experiment. Data- and model-driven approaches have to go hand in hand. However, the alignment of both processes is usually not a smooth homogeneous approach but a more revolutionary shift of paradigms takes place after heavy debates (Kuhn, 1976; Störig, 1988). As an example, the non-dividable liaison was heavily disputed for decades in the last century between the scientific communities of the behaviourism (having a model as a starting point for an observation) and cognitivism (having observations as a starting point for a model). Still many people distinguish both into different disciplines. However, models and data have no difference. Any data is collected by observations and these are structured, even if in a very rough form. No data can ever be collected without a model. Models are therefore part of any data, any observation, any questionnaire (even if the questionnaire is completely open) and any experiment. No model can exist, on the other hand, without data. It will deteriorate to a simple idea. Although both are sides of one coin, data and models were violently separated during the development of cognitive models during the last century. This separation of models and data still persist and is one of the problems for cognitive science meeting current safety problems.

Instead of resolving the contradictions between the different approaches, it was argued whether one or the other approach leads to better models. Debates of this kind usually neglect the primary scientific rule that all approaches claiming to be valid need to be free of contradiction if the object they model is the same. Norman summarized the uncertainty of psychological modelling as follows:

> Each new experimental finding seems to require a new theory (Norman, 1986, p. 125)

Modelling, and cognitive modelling in particular, should be flexible by allowing for integration of additional findings as good as possible. This general requirement is difficult to achieve because the nature of human information processing is complex and difficult to measure. However, other scientific domains deal with similarly complex issues and manage their complexity. An example is the string theory in physics, which allows for additional parameters without the need to revise the entire paradigm.

Requisite Variety

Chomsky (1965) stated that any theory should be adequate for the problem it intends to tackle. Later, Hollnagel (1998) formulated this statement as the principle of *Requisite Variety*:

> The cognitive model should be built in a way that it describes the variety of the aspects in its application field (Hollnagel, 1998)

The principle postulates that it is not required to model the human cognitive system in any detail if one wants to assure that the model itself is correct. A model should depict the main mechanisms but not the full neurological richness of the brain.

The biggest problem of modelling cognition for safety is the scope of what the model should predict. In principle, all human characteristics may become safety relevant, depending on the system architecture and the nature of the process. As discussed in the genesis of existing error models, a variety of issues, in particular the dynamic situational factors, were not assessed or were assessed by cumbersome additions to the high-level structure of the model. Hence, specific error mechanisms or mental demands may be forgotten (e.g. errors due to inappropriate biases; situational awareness).

One therefore has to be careful not to restrict the scope of the cognitive model (or the application field). Limiting the model to the needs of requisite variety will raise the disadvantage that models of human cognition could be limited to the way a system analyst wants the human to behave cognitively (e.g. either focused on decisions or on predicting workload). Due to the increasing integration of different technologies and due to automation, it is not enough to limit a cognitive model to a particular design question if one intends to prevent accidents.

Hence, taking this principle thoroughly into account would mean defining the human information processing itself as application field and not the technology where cognitive errors should be investigated (e.g. aviation, nuclear or medicine). Any other approach would risk having defined the scope of the model too limited for safety.

Computational Effectiveness and Combinatory Explosion

Organisms behave ecologically efficient: Any attempt to build a theory of behaviour that has explanatory power must consider the ecology of nature (Sarbin & Bauley, 1966, p. 175). This position includes that a cognitive model has to subsume experimental findings and neuro-biological capabilities within one approach together with the mathematical calculus within the neuro-biological systems. The three aspects lead to the principle of the *Computational Effectiveness*:

> Any serious model of cognition will have to provide a detailed computational account of how such nontrivial operations can be performed so effectively (Shastri, 1988, p. 336)

This position includes that a cognitive model has to model on a very detailed level, subsuming the entire range of experimental findings, cognitive modelling approaches, and neuro-biological science within one approach.

The biggest problem of achieving computational effectiveness is the problem of the combinatory explosion. Cognitive aspects are highly interrelated and dependent. If one intends to increase the explanatory power of a model by increasing the number of entities to describe cognitive behaviour and cognitive capabilities, these cognitive entities have to be interrelated as well. The effect on how cognitive models develop due to this problem was shown above in the context of the linear model (the classical sequential model of perception, decision and action needed a direct link between perception and action to be able to explain skill-based actions).

Each new element in the model has to be considered in combination with one or several of the existing elements. The number of interrelations therefore goes with the so-called combinatory explosion (also sometimes called the *n over k* problem, because the number of possible combinations is calculated using the operation *n over k*). Any cognitive model attempting to model the interrelations of cognitive entities explicitly runs sooner or later into the problem of a combinatory explosion if going into too much detail (Strube, 1990). The modelling difficulties resulting from the combinatory explosion are extensively discussed in Sträter

(1997/2000). The solution is the concept of dynamic binding, which is an essential principle for an integrated cognitive model.

Law of Uncertainty in Cognitive Modelling

Taking both criteria (requisite variety and computational effectiveness) into consideration leads to the dilemma that any performance model has to deal with uncertainty either with respect to the detail with which it models cognitive performance or with respect to the uncertainty in the validity of the model.

It can be concluded in general that cognitive approaches are more untenable the more they want to predict certain cognitive effects. In accordance the quantum theory in physics (Heisenberg, see e.g. in Kuhn et al., 1976), this phenomenon may be described as a law of uncertainty that results from the complexity of humans' cognitive behaviour:

> The more a certain cognitive function is assumed the more uncertain is the stage
> (within a psychological model) that exhibits this function and, vice versa, the more
> a certain stage is assumed, the more uncertain is the function of this stage (Sträter
> & Bubb, 2003)

In other words, the more correctly a model tries to depict what is going on in the mind, the less general statements can be drawn. The law of uncertainty has only one solution, which is also proposed in this work (as it is done in physics): Describe cognitive processes explicitly, if they are related to the outside world, and describe their functioning implicitly if it is desired to explain a cognitive process. This law of uncertainty should therefore not lead to the conclusion that cognition cannot be predicted at all. However, the uncertainty in cognitive modelling has to be considered appropriately.

Basic Maxims for Cognition

The discussion of the general nature of cognitive modelling suggests abandoning stages of processing because they treat information processing as *Homunculi* (as a distinct number of defined systems with a fixed set of functions). Those homunculi are difficult to be separated and identified in real settings, which leads to the problem that different stages cannot be described as dichotic modules. They are strongly interconnected and have many interrelations, which results in various influences and interrelations between the stages (combinatory explosion). Furthermore, cognition is a reflexive rather than a sequential process. Reflections and reactions of later stages to former ones are usual as shown by simple perception or search tasks (e.g. gestalt laws). Thus, the entire concept of assuming stages of processing seems to be an insufficient simplification of dealing with cognitive processing of humans. A simple extension of the number of processing stages or of the paths between the processing stages improves the predictive power but also increases the uncertainty in the prediction (law of uncertainty). The approach of techno-morphologic stages is tempting from the engineering perspective but not favourable from the cognitive point of view if one considers the iterative nature, the computational effectiveness and the law of uncertainty. In order to establish a proactive approach on cognitive issues in safety, it is desirable to overcome thinking in processing stages. The law of uncertainty suggests not to start thinking in terms of processing stages but to think over the general maxims of information processing.

Cognition is Experience-based

Every cognitive act happens in the memory of the human brain. This memory is built up and based on experience. Even purely mathematical and logical processes need experience at least of the elements used in the logical reasoning (e.g. numbers have to be learned). Experience is not necessarily of external origin. We are, for instance, able to develop knowledge further using mathematical and logical rules and only implicitly using knowledge stemming from external information. This use of experiences can range from unconscious generating of ideas to explicit abstract reasoning.

Experience-based cognition implies that humans tend to gain new experience. The new experience is transferred to an internal representation with the well-known modes of assimilation (of new experiences) and accommodation (inserting the new experience into the existing concepts in memory). New information has to be assimilated before it is accommodated, but the relevance of the new information in the framework of the existing internal world cannot be known before accommodation. Therefore, risk-taking behaviour is a result of the experience-based nature of cognition with a lack of accommodation. Experience-based cognition also explains that humans can argue completely different in different situations since experience is context specific.

Cognition is both Perception-Driven and Goal-Driven

Cognitive performance is linked to the goals a human currently intends to achieve. Goals lead to attitudes one has towards a given situational conditions. Depending on slight deviations of these conditions, completely different cognitive strategies may be used. This link is often ignored in investigating and predicting human error. Classical methods for error prediction, for instance, attempt a prediction by assuming a rather stable set of information processing stages that is used by an operator with more or less success under a given situation (see OECD, 2000).

Cognition is a Process of Dynamic Binding

Human information processing is different from information processing in technological systems. A computer, for instance, has predefined processing stages with a predefined set of functions. Information is processed by information identification, substitution and transportation (Turing, 1936). Anderson (1996) elaborated this approach further by distinguishing between classification, association, combination, and reflection. Reflection is the only process enabling the generation of new contents of memory based on internal thinking processes.

The failed attempt to use the computer-metaphor for modelling cognition suggests that human information processing is fundamentally different. Certainly human information processing has physical structures that make the computer-metaphor a tempting approach for modelling, but the way information is combined and transformed is different. Human information processing is much more flexible than in any current technological system. The human cognitive system creates ad hoc dependencies between cognitive entities. We are able to combine information ad hoc and with respect to very specific situations. This feature is called dynamic binding (Shastri, 1988). Artificial intelligence suggests several approaches of how to model dynamic binding mathematically (Sträter, 1997/2000). An event-based architecture is the one that meets the maxim of being experience-based and avoids the combinatory explosion as well.

Depending on how well the current binding fits the external situation, different types of dynamic binding are implied and different control modes are necessary (Hollnagel, 1998).

Cognition needs Cognitive Dissonance

Cognition needs some kind of energy to proceed. Festinger (1957) named this energy *Cognitive Dissonance*. Dissonance can be understood as a mismatch between the externally required and goal-driven desired cognitive coupling. For instance, dissonance may occur if a task requires an operator to be monitory but the operator intends to be active in an emergency situation. Cognitive dissonance is related to the concept of mismatch activity of orientation reactions in neuro-psychological research (e.g. research on EEG; e.g. Sommer et al., 1998). Cognitive dissonance depends on two parameters: the situational aspects and the cognitive abilities.

A mismatch is a necessary but not a sufficient precondition for conscious processing. In many situations we are able to cope with situations on a sub-conscious level, even though dissonance is present. This is the case if the situation can be matched to some learned experience without the need of adaptation to the situation. In cases where the situation cannot be matched successfully, a higher level of cognitive behaviour is required. According to Rasmussen (1986) and Reason (1990), the levels of behaviour range from skill-based via rule-based to knowledge-based behaviour. According to the theory of cognitive dissonance, the skill-, rule- and knowledge-based levels have mechanisms, properties and a location in common. The properties include signal-noise thresholds and a certain amount of sluggishness while moving from one level of behaviour to another.

Cognition means Reduction of Cognitive Dissonance

In cases of mismatch, human beings try to cope with this difference by reducing it (Festinger, 1957). This means humans try to impose stability on their internal world after a mismatch has been experienced. It is important to note that this does not imply a stability of the external world. This aspect of cognitive processing is a decisive maxim. Even if there is a technical disturbance, we can operate a plant, drive a car or fly a plane only if we are in internal stability. Situational aspects that endanger the stability are ignored or even rejected (e.g., counter-cues, side-information). This mechanism, called confirmation bias, is in full accordance with the theory of epistemology: we assume something as true, if it fits into our internal world without any contradiction (Keller, 1990).

Cognition has Binary and Unspecific Mechanisms

The cognitive system is not a binary system as often stated in the literature. The neurological nature would suggest this from the way neurological cells generate their response (potential threshold). However, the neurological system of the brain has an additional means to control the activation of cells and the spreading of activation, namely by using chemical substances (neurotransmitters). The neurological system is hence a tertiary system, not a binary one. The neurological as well as the neurotransmitter-related means of control can be positive (reinforcing) and negative (inhibiting).

Unspecific mechanisms are essential for understanding general modes of processing, like certain emotional states, motivation, anger or fear, which can all play a role during decision-making in critical situations.

Characteristics of the Cognitive Processing Loop

The Cognitive Mill

Stages of information processing are doubtful and the law of uncertainty prohibits modelling cognition referred to certain cognitive facilities or homunculi (e.g. memory, perception etc), because they lead to doubtful predictions. The principal cognitive processing loop suggests overcoming the limitations of the techno-morphologic information processing by describing cognitive abilities rather than cognitive processing modes. The basis of the functions the cognitive apparatus may perform is the distinction of the internal world and the external world. This can be derived from philosophy (e.g. Störig, 1988) as well as psychology (cf. Kieras & Polson, 1985). Katzenberger (1967) interpreted subjectivity as a result of seeing the external world from the internal perspective. Popper (1997) distinguished three worlds: the physical world, the internal world and the conscious world. He further stated that the conscious world is implied by the physical and the internal world.

- The external world is the set of objectively given information, which comprises all information external to the cognitive system. Objectively given information may cue the cognitive system (cf. Sanders, 1975).
- The internal world is the set of subjective or internally given information, which includes all the information internal in the cognitive system. It is sometimes called the mind-set (cf. Husserl, 1976) or mental model (cf.

Gentner & Stevens, 1983). Subjective given information is related to the declarative knowledge of a human being (e.g. Mandl & Spada, 1988) but also to attitudes, wishes, goals etc.

The human brain permanently collects information and compares it with the internal representations about the world. This is sometimes conscious but most of the time it is an unconscious process. Neisser (1976) called this process the *Cognitive Mill*. The metaphor can be used to describe the way in which the external world and internal world are aligned. Objectively given information and internally given information are continuously updating each other via actions or stimuli.

The result of this process is a 'recognize-act-cycle' or 'perceptual cycle' (Card et al., 1983) as represented in Figure 3.2. The cognitive mill is the basis for accommodative performance. If the internal and external worlds fit well to each other, no accommodation is required and the cognitive system is in an equilibrium state (Festinger, 1957). Control interventions may be due to a stimulus-driven 'disturbance' of the internal world or changes in the external world due to human actions. The recognition of a situational pattern does not imply an abnormality in the human brain per se. Without disturbance, the human will act on a subconscious or low-conscious level, or a skill-based level (e.g. walking until we stumble). If the recognition does imply an abnormality, the conscious processing will be involved for this abnormality. This so-called *Cognitive Dissonance* is a mismatch between learned behaviour and the actual situation. Dissonance is implicitly represented in the decision points for a higher-level cognitive performance in the GEMS-model. This means, on the other hand, that the level of cognitive behaviour is directly related to the amount of cognitive dissonance.

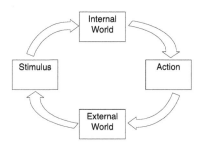

Figure 3.2   The cognitive mill (e.g. Neisser, 1976)

The principle processing loop implies that a cognitive act somehow relates both worlds to each other. Relations are linked to the concept of procedural knowledge (Mandl & Spada, 1988) or the concept of *Running the Mental Model* (cf. Gentner & Stevens, 1983). The link established may induce a delta of energy between both, which leads to dissonance. In such cases, ad hoc binding of the external and internal worlds is established in order to minimize the difference and level out in a new equilibrium state. The levelling may be performed in simple associations up to several iterative steps of minimizing the difference as usually required in more complex situations. The outcome is an established relation between the internal and external worlds. Overall, each complex cognitive process for aligning the external and the internal worlds contains three general cognitive acts before an action is executed:

- cognitive coupling,
- cognitive binding,
- cognitive levelling.

Central Cognitive Acts

It was discussed in the context of the GEMS-model that three basic cognitive error types could be observed:

- information-driven aspects of human errors (like perceptual confusion and inattention, information overload, complexity of the problem),
- aspects related to the resolving mechanisms for the alignment of information and goals and for finding a way forward (interference, rule strength, biases in reviewing information and goals, stereotypes), as well as
- top-down related aspects of generating actions (like repetition and stereotype takeover, generality of rules, overconfidence).

These rather generic error types are independent of the different levels of cognitive behaviour, and manifest themselves in simple signal response behaviour as well as rule-based actions or complex knowledge-based behaviour. It seems that the error mechanisms are not a feature of the specific cognitive level but more of the same origin. The only way to explain the common nature and take into account the rule of simplicity of nature is to assume that they are a result of central cognitive acts.

Central cognitive acts are derived from the relation of the external world and the internal world as described above. Mismatch is related to cognitive coupling, resolving mechanisms are related to cognitive binding, and cognitive levelling seems to be related to the generation of actions. The basic cognitive acts are generic and may lead to similar errors independently from the cognitive level (skill-, rule-, knowledge-based). In turn, the cognitive levels as used in the skill-, rule- and knowledge-based approach reflect these generic acts rather than cognitive entities.

However, the central cognitive acts should not be understood as a central processor, because the techno-morphologic picture of the central processor would understand the remaining parts of the cognitive system (e.g. memory) only as passive resources of information and not as active parts of the information processing.

The central cognitive acts are deciding whether an equilibrant state has been achieved or not. The concept of the central cognitive acts is compatible with the skill-, rule- and knowledge-based classification of behaviour but simplifies the architecture to overcome the limitations and unnecessary complexity regarding the way the cognitive system shifts between the different levels. The rhombic shapes in Figure 2.5 represent the decision processes to move into the next higher level of cognitive behaviour. The approach of central cognitive acts assumes that all three decision processes are handled by the central cognitive acts. The concept therefore keeps the experiences and scientific background of the GEMS-model but simplifies the architecture regarding the decision elements by postulating only one instead of three different resolving mechanisms.

The simplification is not only rationalized by the law of simplicity of nature. As described in the following chapter, this approach also allows including those aspects that were difficult or cumbersome to include into the GEMS-model, such as the interaction of knowledge-based aspects and skill-based behaviour or the impact of dynamic contextual conditions on cognition. Herewith, the principle of central cognitive acts is a true further development of the ladder-model. It keeps the scientific experiences reflected in the ladder-model but includes additional explanatory power.

The Connectionism Nature of the Internal World

The brain consists of a complex net of cells. The modelling corresponding to the nature of the human brain is the so-called *Connectionism* (e.g. McClelland & Rummelhart, 1981; Kempke, 1988). Compared with classical block-models, the connectionism approach is a far more realistic assumption for the architecture of the brain.

A connectionism approach that represents the complexity of human cognitive processes can cope with the dilemma of uncertainty (i.e., that no cognitive model can be certain). It explains the relationship of cognitive processes to human performance, its interdependencies, and the relationships to contextual and situational conditions as an experienced-based network. The connectionism representation is hence able to represent the situational conditions, under which behaviour is performed, as well as the causes for possible errors.

The connectionism approach is able to represent two important properties of artificial intelligence: polymorphism (that the same semantic information can be interpreted differently in different contexts) and encapsulation (that specific semantic information can be processed based on various experiences). Both are essential to understand human error. The connectionism approach allows modelling the influence of the dynamic situational context.

The connectionism approach models cognitive processes using the metaphor of the neural cells, which have the following characteristics: (1) axons and dendrites for information input and output, (2) a level of activation triggering other cells if it exceeds a certain threshold, (3) activation-processing through the net of cells, and (4) learning by frequency of use of cell-connections (Hebb-Law). Based on these characteristics, the status of the internal world is represented by the structure of the neural net. It consists of cell-connections, possible ways of activation (i.e., the knowledge for processing), their current activation (i.e., the status of the internal world), and the activation spreading through the structure of the cell-units (i.e., the information processed within the structure of the cell-units). As pointed out in the next chapter, these properties are essential to describe how the central cognitive acts are realized and are necessary to understand how human errors are generated.

The Cognitive Loop of Processing Information

The cognitive system is a cognitive mill with central cognitive acts to level out the dissonance between externally and internally given information. Figure 3.3 illustrates how the cognitive coupling and the connectionism representation are related to generate cognitive behaviour.

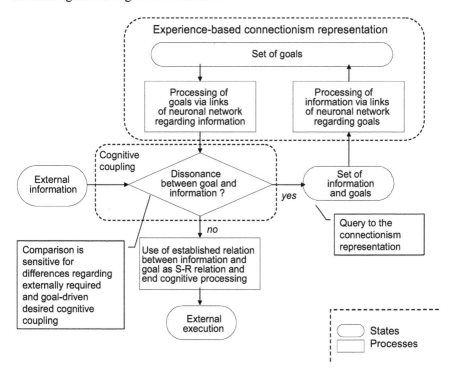

Figure 3.3 **The cognitive processing loop – coherency between coupling processes and experience-based knowledge**

In the connectionism approach followed here, dissonance is operationalised as a mismatch between the external set of information and internal set of information. This approach understands terms like 'task' or 'goal' as context dependent concepts. External information entering the cognitive loop becomes a task (an alarm, a prescribed list of action etc). The task may be transferred into a goal after being processed into the experience-based connectionism representation of the internal world (supposing it is not lost during the evaluation). The information remains the same but the way it is used in the cognitive system changes. Whether information is to be understood as a task or a goal depends on the part of the processing loop where it is used, but it is not an inherent property of the information itself.

External information is processed in the context of the given internal information. In case there is no dissonance between both, the established relation is used and an action is performed (i.e., skill-based cognitive behaviour). In case the external information mismatches with the internal information, no relation can be established and dissonance is produced. This dissonance triggers the further evaluation of the mismatch. The experience-based connectionism representation is used to find a new set of information that closes the gap between internal and external information. If this is found the action is performed (i.e., rule-based cognitive behaviour). In case the gap can still not be closed, further loops may be required to find a relation between external and internal information (i.e., subsequent knowledge-based attempts to close the gap).

The use of the ladder-model to explain the cognitive loops shows that the cognitive loop of processing information is fully compatible with the resilient aspects of the techno-morphologic information processing models. It will be shown in the next chapter that this approach is also able to overcome the limitations of the techno-morphologic models. The connectionism representation allows integrating the dynamic situational conditions, under which behaviour is performed and which may cause errors. The central comparison explains the differences in decision-making on the different cognitive levels and the interdependencies of the skill-based and knowledge-based level (e.g. biases, heurisms). The comparison of internal and external information is sensitive for differences regarding the nature of the externally required and goal-driven desired cognitive coupling. This element of the model is essential: if a human is involved in a dynamic task for instance, a static task cannot be performed at the same time. In such a case the human needs to decide on one or the other or will require time to get a bigger picture of the information. This basic assumption also has a considerable link to the cognitive streaming theory (Jones & Farmer, 2001). The theory states that, given a human is busy with a serialization task (e.g. counting), he is no more able to serialize other information, even if serialization is required on a different information channel (visual vs. auditory) or in a different modality (spatial vs. temporal). It will be shown in Part III that this phenomenon is not restricted to control tasks but also holds for communication within groups.

Chapter 4

Mechanisms of Cognitive Performance and Error

Modelling cognitive processes is a scientific process. As the genesis of modelling showed, cognitive modelling always depends on an actual current scientific 'wave' or 'fashion' of cognitive, technological or social developments (cf. also Kuhn, 1976). The rules for scientific reasoning can be used to find a way to approach the complexity of the issue.

Three basic criteria for modelling are objectivity, validity and reliability. Objectivity describes that the model should be independent from the perspective of an observer. This criterion is not empirically provable but inferable by assuring that the model reflects different perspectives. Validity means that the model should reflect cognitive processes, which is provable by comparison of the model predictions of human behaviour (e.g. in incidents or accidents). Reliability indicates whether the predictions are stable, what is provable by comparison of predictions over time or application fields.

Applying all three criteria to cognitive modelling creates a cognitive model that is sufficiently objective, valid and reliable enough to include cognitive processes into safety assessments. It has to be elaborated, how techno-morphologic models fit into this new view but also how to better integrate the issues of dynamic situational context, the errors of commission and the interrelation of workload and cognitive performance. An integrated model will not result into a completely new model but into a coherent view on the variety of existing approaches. It enables an easier combination of the views of error modelling, cognitive design and organization of workplaces.

Cognitive models have roots in different scientific domains like philosophy, mathematics, cybernetics, medicine or neurobiology, psychology, or physics and also contain various aspects of many other sciences. The different domains can be complied in three main perspectives. The modelling perspectives are:

- The physiological perspective, which will provide a rationale for the general structure of the cognitive processing loop and its link to emotional states.
- The mathematical perspective, which will provide a rationale for the computational effectivenss of cognitive binding, the experience-based architecture of memory and the generic decision-making mechanisms.
- The information theoretical perspective, which will provide the rationale for the memory-span, characteristics of cognitive control modes, logical reasoning and reaction times.

The discussion of these perspectives contributes to the objectivity and validity of the approach of the processing loop. The consideration of all perspectives will complement the current way of approaching cognitive modelling. The perspectives will be discussed in the light of the generic scientific rules mentioned in the introduction. The principle of simplicity of nature will be discussed mainly from the psychological perspective. The rules of scientific logic to assure traceability will be discussed in the context of the mathematical modelling and the information theoretical perspective revealing an approach that is inherently logical and free from contradictions (i.e., avoiding the inquisitory logic). The rule that a model should have more explanatory power than the preceding ones will be discussed in the context of the information-theoretical perspective and the psychological perspective.

Overall, the chapter will show that the cognitive processing loop approach is not an entirely new model but is complementing and integrating existing models and approaches that have been used isolated from each other so far. Complementary aspects included are cognitive resolving mechanisms, attitudes, emotional states and decision-making as well as their interrelations. It claims the importance of such an integrated approach for a proactive integration of cognitive aspects into the design of work-processes and a preventive strategy of safety management.

The Physiological Root of the Processing Loop

The insight of physiological findings to cognition has developed well in the last decade. The related findings may improve also the view on modelling cognition. A more realistic approach of human information processing may be achieved, by integrating this perspective into cognitive modelling. A cognitive model should certainly not reflect every single neuron of the brain but also should not contradict the findings from physiology and should take into account the principal findings in this area regarding the neurological structures and the neurological processes.

The Piece of Interest

Not all aspects relevant for modelling cognition can be elaborated here. However, some important ones should be highlighted.

If one intends to model a human's cognition in a certain situation, he or she has to make a prediction about the processes in the brain at the end of the day. Although making considerable progress during the last ten years, even the intensive research on the different fields of brain-research (medicine, biology, neuro-science, psychology, and philosophy) were not able to give a reasonable and sufficient approach to explain or model how human cognition is linked to the brain functions. However, the results available give important hints on the missing bits in achieving an integrated model. Physiological knowledge can be used for falsifying (or validating) certain cognitive approaches and allows herewith to open the view on an integrated model. The physical system enabling human beings to perform their cognitive behaviour is the human brain and has a highly complex nature. Figure 4.1 gives an impression of the architecture of the brain.

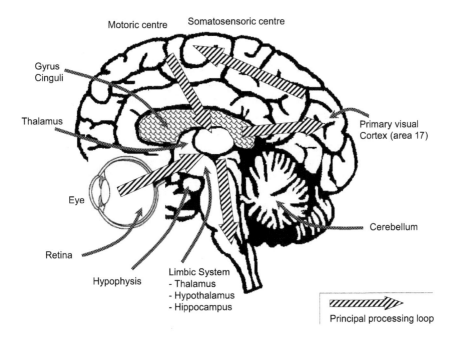

Figure 4.1   The piece of interest – the human brain

The brain consists of a huge amount of neural-cells (>> 10 billion cells) that are highly connected with each other and may be differentiated into major areas like the motoric centre, visual centre. Although some areas have been found to be more responsible for certain cognitive processes (e.g. motoric centre, see-cortex), the connectivity of the cells and the role of the different areas is (un)fortunately not of a deterministic nature.

One essential problem that was not solved within brain-research is to link the physiological findings to the possible cognitive performance of the brain. One reason why it is so difficult to link cognitive modelling with the brain's anatomy is the mismatch of the structural models with the general architecture of the brain. Certain research results were only able to explain a small part of the spectrum of the performance of the brain by using partial models. According to the law of uncertainty, this reduction into partial models of human behaviour always causes problems in applying them to reality. On the other hand, a one-to-one representation will not be of interest for the engineering of system safety, as it would be too detailed.

Despite this problem, the figure indicates a clear mismatch between the architecture of human cognition as modelled within the techno-morphologic approaches and the neural architecture. Physiological investigations showed, for instance, that seeing is not an isolated functional entity of a visual centre but that visual performance virtually interrelates to all brain regions. Applying the rules of scientific logic, this is a main reason why a simple stage-based approach cannot be true. A perceptual stage is not in accordance with the physiological knowledge and hence must be wrong. Two further important aspects of physiology should be discussed as they support cognitive modelling. Various areas exist that release a set of different neuro-transmitters with a broader impact on a more or less specifiable number of cells (e.g. alertness, mood).

The processing loop as illustrated in Figure 3.3 of the previous chapter is compliant with the general architecture of the brain (the processing loop is indicated by the dashed arrows in Figure 4.1). The paths in the processing loop match the general path of information processing in the brain. It is no coincidence that this path is the optimal one if one takes the principle of simplicity of nature into account. The general brain architecture assures short and efficient distances to process external information into actions on the shortest path possible. Any techno-morphologic approach would require a much longer path and hence more neural structures and more energy to process information. As nature usually takes the most efficient way for a mechanism, it must certainly also be true for the processes of the piece of interest.

Mechsner (1998, p. 70f) found that consciousness is related to the limbic system and made the hypotheses that consciousness started with the regulation of emotional aspects as basic resolving mechanisms. Consciousness always comes into play if mismatch occurs and it is no more guaranteed that the equilibrium state can deal with a situation. Cognition has to become more creative and consciousness is used to emphasize certain aspects while inhibiting others. In other words, consciousness comes into existence with change (similar to the way electrical power is generated by a move of an electron through a magnetic field). The change is the information processing activity. The strong link of emotions and resolving mechanisms affects the information gathering and selection abilities of human information processing (e.g. during periods of fear, anger, alertness or bad mood).

Neurological Issues

Psycho-physiological measurement Psycho-physiological measurement developed the EEG (Electroencephalogram) as a means to measure brain activity during cognitive tasks. Several cognitive aspects could be linked to the activity of the brain so far and many additional experiences are to be expected. An extensive discussion can be found e.g. in Sommer et al. (1998). The main measures to highlight in the context of error modelling are the P300 and the N400 components (GUIRE, 2004).

Physiological aspects (like the physiological measures of the P300 and N400 component for instance; see below) will be used next to the criteria of computational effectiveness to elaborate the integrated model. The P300 is an electrically positive wave appearing about 300 milliseconds (ms) after the onset of a significant event. The P300 amplitude increases as events become less frequent (in the words of the ladder-model, if they become more rule-based). If perceptual processes consume more time, the latency of the wave increases. The P300 effect seems to underline the hypotheses that conscious control is expressed by suppressed inhibitory signals (Milner & Michalski, 2003) and that dissonance is represented by a positive activation wave. The N400 is an electrically negative wave appearing about 400 ms after the onset of a visually or acoustically presented word. The N400 amplitude increases when the presented word does not fit into the semantic context, which has been established by previous stimuli.

Both P300 and N400 are hints for the central cognitive act of levelling out the mismatches between external world and internal world. They disappear as soon as an established relation is achieved. The difference in timing between P300 and N400 is 100 ms. This is in the range identified in cognitive ergonomics. Card et al. (1983) suggest that the time of a single iteration in the processing cycle is about 70 ms (ranging from about 25 to 170 ms).

Other hints on the role of certain neuronal structures for cognitive modelling may be obtained from research on psycho-physiological deficiencies. As an example, the so-called pyramid-cells connect the cortex area with the motor system. These cells usually have a vertical alignment to feed information back from the cortex properly. If the cells are disordered (e.g. more horizontally aligned) there is no coordinated processing possible (Comer, 2001).

Cyclic nature of neurological processing The human cognitive system is parallel and sequential. However, lining up the neurons in a sequential manner would lead to a maximum of about 100 neurons in a row for any, even complex, information processes (if no iterations or cycles are assumed). This fact, sometimes called *100-Step Rule*, is another hint that the information processing is organized in a cyclic way rather than sequentially. In EEG, this cyclic nature can be observed as different frequencies are usually distinguished into alpha, beta and gamma waves of cognitive activities. The alpha wave has a frequency of about ~10 Hz (10 cycles per second). The frequency of the alpha wave increases with increased demand of cognitive activity (Malsburg, 1986, Krüger, 1998). The frequency of the alpha wave is highly correlated with the reaction time and memory span (Shearer et al., 2004). If the frequency of the alpha wave is reduced, the reaction time is reduced and the memory span (short-term memory) decreases as well. One well-known effect is the reduction of the memory span and of the frequency of the alpha wave of older people, where both reductions are proportional to each other.

Cognitive dissonance and consciousness Cognitive processing consists of conscious and unconscious processes. Conscious processes are self-reflective and controlled, unconscious processes uncontrolled and without an opportunity to reflect them (to look at them with the inner eye). This difference, investigated in attentional research (Wickens, 1984; Sanders, 1983; Posner & Snyder, 1975), was one reason to assume shortcuts of certain cognitive stages. However, the fact that biases and heurisms exist in skill-based actions suggests that conscious and unconscious processes are not done by distinct stages but are different ways of using the same information. Conscious aspects are influenced much more by unconscious processes, as we can be aware of (of course, as per definition). Search and recognition of faces is performed by unconscious processes, and the own name effect is known as a proof of how important unconscious processing is for our cognitive system. Such effects are not only clear counter-arguments against sequential shortcuts but also implicate that conscious and unconscious processes are happening within the same system.

The central cognitive act of levelling suggests an easy but efficient way to distinguish both. Once information from the external world causes dissonance with the current internal set of information, mismatch is produced. Mismatch means energy. The mismatch consequently is measurable as the P300 component in the EEG. The mismatch itself is already the trigger for consciousness and not only an event that triggers some cognitive institution to launch consciousness (as represented in the ladder-model). Consciousness means 'pulling' the brain out of the equilibrium state. The amount of how much it will be pulled out depends on the amount of mismatch. The memory content can principally be transferred from unconscious to conscious processing if the energy of the cognitive dissonance is changed. The mathematical perspective will show how semantic, explicit information processing can be explained by the use of mismatch.

Reflexive cognitive processing, control modes and levelling Reflexive cognitive processing and control modes were discussed as being essential to understanding the active aspect of human behaviour that sometimes leads to errors of commission. In order to understand the mechanisms behind this, it has to be understood how aspects in a given situation are processed in the cognitive control loop.

Cognitive processes are reflexive. Humans can reflect their own content of memory and make logic operations on the content of memory. Control modes were discussed in the previous chapter as being a set of processes representing this reflective cognitive behaviour. Recent brain research (Münte, 2002) showed that the control modes could be allocated to the region of gyrus cinguli. This brain area is located between the cortex and mid-brain and between left and right hemispheres. It could be shown that the area gets more active as soon as mismatch between internal memory content and external information (between goals and action) occurs. It also gets active in the case of lies, or slows down the reaction time in the case of mismatches or error recognition and serves for the alignment of different memory contents (e.g. resolving of conflicts in memory). The latter function of memory alignment is known from compatibility research in ergonomics.

This part of the brain is apparently somewhat needed to deal with mismatch, to enable detailed processing in the cortex area and to level out cognitive acts. The ability to react on specific aspects of a situation before knowing what to do in detail (the so-called orientation-reaction) is the strongest evidence for assuming that the mismatch is pre-processed before detailed cognitive processes start. The mathematical perspective and the perspective of information theory will discuss how this part of the brain is essential for understanding memory span, learning and reasoning.

Endocrine Issues

The tertiary nature of human cognition   The techno-morphologic paradigms preferred to interpret neuronal cells as binary, digital elements. This view, appealing because it supports the idea that the computer becomes an approach for modelling, is not able to represent all processes in the brain. The functions are heavily influenced by the chemical, endocrine system as a third dimension of cognitive processing. This third dimension provides a general trigger to the emotional component of cognitive processing. This emotional component is also known to support or hinder the memorization of external information and cognitive performance in critical situations. The limbic system (consisting of the hippocampus, hypothalamus and thalamus region) is known as the system creating emotions and influencing the performance of short-term memory. The thalamus region is often placed in the role of a switchboard between sensory input and cognition. Its role is, however, far more essential that just switching. Nature would not create a complex structure that is just responsible for information forwarding.

Again, the energy produced by the cognitive dissonance seems to trigger the third dimension of cognitive processing. Cognitive dissonance triggers the emanation of endocrine substances into the limbic system and herewith a row of several cognitive aspects is controlled. It is known from serious disasters and accidents that the people involved decouple from the external world after the accident and suffer from serious emotional problems. This well-known effect of self-protection enables the cognitive system to cope with mismatches of the internal and external world that cannot be solved. Herewith the endocrine system supports the cognitive levelling and can control levelling.

The endocrine system sets general and specific activation levels in the processing loop. The subsequent section dealing with the mathematical perspective will elaborate how the endocrine and the cognitive system are interrelated. Important effects of human behaviour impacted by the endocrine dimension are the following.

- The 'sharpness' of conscious processing is controlled depending on the deviance of the internal and external world. Humans can focus attention on a particular aspect of a situation that can then be elaborated in detailed or distribute attention amongst several items that only then can be processed in an abstract way. This property, sometime called the *Searchlight Metaphor*, can be explained by changes in the threshold level introduced into the cognitive part of the cognitive system.
- Related to the effect of focused and distributed attention is the interdependency of cognition and stress. Tunnel vision, the reduction of the memory span coupled with emotional effects, is a known effect from critical accidents (e.g. from voice recorders of pilots shortly before an accident).

- Achieving the relation of cognition to vigilance, sleep, dreaming and alertness (cf. the ARAS, Ascending Reticular Activation System).
- The interrelation of cognition and emotional states combined with critical cognitive processes (such as frustration, enthusiasm), which lead to cognitive effects of overtrust or undertrust.

Relationship of Cognitive Aspects to Organic Regions in the Brain

The cognitive system is a highly connectionist and interdependent system. In the parallel distributed processing approaches it was discussed that a certain cognitive content (e.g. a picture of a grandmother) is either distributed in the neural net or allocated to a particular cell. Recent brain research found that the cognitive system is neither of both extremes but both distributed and with specific functions allocated in certain areas of the brain.

The allocation of cognitive characteristics and neural structures as discussed in this chapter are summarized in Table 4.1 according to Schmidt & Thews (1987). Only the general architecture is reflected in order to help understanding the interactions of different cognitive processes, which are often not considered in error modelling.

Table 4.1   General allocation of cognitive characteristics and neural structures

Neural correlate	Cognitive function	Cognitive effects
Signal sensors	Pre-processing of signals	Signals arranged according to space and time
Hyper-complex cells in the visual processing area 17	Image store	Spatial perception of 2 and 2.5 dimensions
Limbic system, Thalamus region	Mismatch negativity (from EEG research)	General comparison, ranking judgement (earlier, later, bigger, smaller, higher, lower), link to emotions and stress, alignment with vestibular-sense
Gyrus cinguli, Hippocampus area	Central cognitive acts (coupling, binding, levelling), Control of consciousness and attentional resources	Pattern matching, Bounded rationality, dissonance, cognitive tendencies, link of certain perception to emotions, cognitive control modes, reflection
Cortex	Semantic processing	Semantic distance, experiences, biases, heurisms
Cerebellum, motor system	Execution	Fine tuning of movements, spontaneous reactions
Alpha frequency	Pacemaker for the processing cycle	Variety in memory span

Behaviour as a Result of Cognitive Binding

Dynamic Binding and the Generic Architecture of Memory

The effect of combinatory explosion Within the cognitive processing loop, the internal world is represented in the experience-based connectionism representation. To come to an idea of how this experience-based connectionism representation is constructed, mathematical rules may help. The essential key-problem of finding an appropriate modelling approach is the huge variety of possible performance of the human being and the extensive interaction of many pieces of information and the parallel processing information.

According to the law of uncertainty, the functional blocks defined in classical models will only be able to describe a comparable small set of cognitive performance. They, for instance, are not able to represent the influence of dynamic situational factors on the cognitive processes. Making the model reflect these dynamic situational factors can only be achieved by distributing certain functions across the functional blocks used in the model. This harms the inherent logic of the model.

The most important mathematical rule for describing this modelling effect is the rule of combinatory explosion (e.g. Kempke, 1988). A simple example may show the problem of combinatory explosion for cognitive modelling: it is known that plants produce oxygen via photosynthesis and it is known that animals and human beings consume the oxygen produced (Figure 4.2).

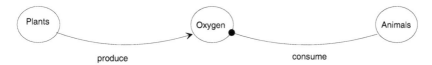

produce   consume

Figure 4.2   Representation of the example of Rosenfeld using bilateral binding (from Dörner, 1997, p. 110)

The figure represents the obvious, daylight truth of the relationship. It is also known that plants use oxygen in the night. A classical modelling approach would only be able to represent this fact by introducing an additional node to distinct day and night. To represent the logic correctly, two additional conditions also need to be introduced to represent the different aspects of plants producing oxygen and plants consuming oxygen depending on the condition. The resulting Figure 4.3 is the only possible solution of modelling the additional aspect properly, if one restricts modelling to the techno-morphologic restriction of bi-polar relations.

In the next step, the model may be refined further so that different types of plants are distinguished. With each type of plant introduced, the number of nodes required is increasing by the number of existing nodes, which are interrelated to the new aspect. The higher the interdependency becomes the more nodes to be added. This effect goes with the mathematical expression of *n over k*. Beside the explosion of nodes to be added, the cognitive system would also be required to shift the meaning of the nodes from formerly one node representing the concept of plants into three nodes (plants, plants under the condition of daylight, plants under the condition of night). Each new aspect would further distribute the semantic meaning into the nodes added and would make the access to the information of what a plant is and does increasingly difficult. This distribution of information is called polymorphism. The fact that the semantic meaning of the term 'plant' remains, although it is used in different contexts, is called *Encapsulation*.

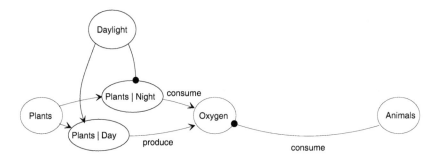

Figure 4.3   Representation of one additional aspect of truth into the example of Rosenfeld (from Dörner, 1997, p. 110)

To model the enormous amount of information humans are capable of processing using bilateral binding is not problematic if many parameters are interacting and are interdependent under particular circumstances. Shastri (1988) discussed these problems in detail. Any approach using bilateral binding (called 'Socrates inference' by Shastri) ends up in the question of how the binding is created by the cognitive apparatus. What are the data used? This holds even for language production systems and artificial intelligence approaches. Who decides to split and how to split concepts?

The problems (like combinatory explosion, encapsulation and polymorph relations) are discussed in detail in Sträter (1997/2000). Additionally to the problems of combinatory explosion, encapsulation and polymorph relations, three aspects are not considered in bilateral binding approaches:

•   The generation process of how relations between concepts are constructed.
•   The use of the relations in certain situations.
•   The calculus of how the relations are mathematically calculated given a certain external world.

As discussed in the context of the dynamic situational factors, humans are able to behave in a very different way in two situations that are similar in many respects but different in only one specific effect. For instance, the controller in the mid-air collision probably would not have committed the error of intervening into the actions on the airborne side if he had not been working on a delayed third aircraft, although the rest of the situational conditions would have remained exactly the same. A model not taking into account the shift in reactions based on slight differences or the dynamics in the situation would not be able to represent the cognitive processes and the information considered in the processes properly. Any model attempting to represent those slight deviations needs to overcome the problem of the combinatory explosion.

The problem of the combinatory explosion is central to understanding how the structures in the brain generate human behaviour differently for two situations that only differ in one detail (Shastri, 1988). Mathematically speaking, this needs an approach that is able to deal with the effect of encapsulation and polymorphism.

Connectionism approach of information processing The last section pointed out that a reduced modelling of human behaviour in functional blocks always causes problems in applying these models to real human behaviour. In principal, these models are only able to explain a small part of the spectrum of the performance of the brain, because they run into the problem of the combinatory explosion. According to the law of uncertainty, therefore, any of such approaches to model human cognition is, in principle, insufficient to model the huge spectrum of possible human performance. A good starting point to deal with the effect of combinatory explosion may be a combined approach, which integrates cognitive modelling, experimental, and neuro-physiologic aspects. Such combined approaches are discussed for some time as valuable approaches for modelling cognition under the terms *Parallel Distributed Processing* or *Neuronal Nets*. They were often discussed as a new paradigm for cognitive error modelling (e.g. Neumann, 1992). These approaches are trying to model cognition based on neurological findings about the brain (e.g. Stoffer, 1989) and shift from the techno-morphologic approach into a neuro-morphologic approach, which apparently fits better to the nature of human cognition. They are not forced to make any assumptions about certain information processing stages and are concentrating on being in accordance and close relationship to the architecture of the brain. The idea to model cognition using the neurological paradigm is not new. Thorndike (1922) introduced the term as an approach for a general theory of mathematically describing learning. Neuro-morphologic models were already developed in the 1950's and used to explain basic neurological functions like pattern recognition (Hebb, 1949). They were revised again in the 1990's in the context of artificial intelligence and expert-systems (e.g. Rummelhart & McClelland, 1986; Kohonen, 1988, Nauck et al., 1994).

The approach was initially considered as a suitable algorithm for the recognition and processing of visual or auditory patterns (image processing, speech identification) such as the word recognition model (McClelland & Rummelhart, 1981) or the model for the typo errors (Rummelhart & Norman, 1982). The results from the initial applications were so promising that this approach was also considered as an alternative to modelling higher information processing capabilities such as decision-making (Neumann, 1992; Strube, 1990; Stoffer, 1989). Approaches to cognition psychology modelling can be found among others in Shastri (1988), Fu (1993) or Emmanji et al. (1992). Compared with the stage-based approaches of modelling cognition, connectionism imposes the least restrictions regarding the structure of the information processing mechanism. The spread of activation is controlled by the network itself. The evaluation of the concept and experience do not require a central processor. A network is capable of dealing with variations of patterns. Taking into account this high adaptability, this approach seems suitable for solving problems that the models discussed so far encounter.

The view on cognition in the neuronal nets approach is quite distinct from the view in the conventional approaches of assuming processing stages. Cognitive performance is modelled in more detail and higher-level processes are seen as the 'macro-level' of the 'micro-level' processes (McClelland & Rummelhart, 1981). Herewith cognitive phenomena may be explained that cannot be described within the conventional approaches, like the semantic priming effect, associative recall or learning of behavioural patterns (Fitts law). Nevertheless, problems occur if one intends to go beyond modelling simple associative information processing. Explicit semantic reasoning and understanding cannot be explained. This holds even for simple reasoning of predicates like logical inferences (cf. Puppe, 1988) or text understanding (Schnotz, 1988). In conventional approaches, those abilities are usually covered by decision theoretical approaches (e.g. Dörner, 1976). The approach also failed so far to include the results of modelling the memory span (Miller, 1956; Sanders, 1975; Attneave, 1974).

The promising advantage of having the potential to overcome the restrictions of combinatory explosion and the law of uncertainty could therefore not be used for modelling human behaviour. As a result, many so-called hybrid models were established, like the SOAR approaches on unified theories of cognition (Newell, 1990) or the model PSI (Dörner, 1999). These approaches are modelling cognition by using different sub-models for different domains where predictions have to be made. For instance, neural networks are used for pattern matching and eye tracking, while explicit stage-based approaches are used for decision-making. The sub-models represent an empirically built collection of models but are not necessarily fully compatible to each other. This makes it difficult to use the models for human error modelling, as only an integrated model is able to represent the complex interaction of the different human characteristics under dynamic situational conditions.

The principle of dynamic binding of information Hybrid models are one possibility to deal with the problem of the combinatory explosion. However, they do not overcome the problem. Encapsulation and polymorphism are achieved by using a collection of approaches but are not modelled in an integrated manner. In order to solve the coding and retrieval problem, the literature on artificial intelligence discusses the object-oriented approach, fuzzy sets, probabilistic networks, as well as neuronal networks. These various approaches and their limitations regarding the potential to deal with combinatory explosion, encapsulation and polymorphism are discussed in detail in Sträter (1997/2000).

The solution is to understand a cognitive act as an ad-hoc built relation between cognitive concepts rather than an information flow in a predefined relation of functional blocks that 'transform' or 'transport' the concepts from perception to decision-making and action. Figure 4.4 reflects this generic thought of dynamic binding of information. Dynamic binding was proven as existent in physiological investigations (Malsburg, 1986).

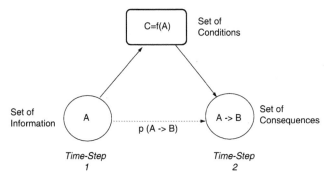

Figure 4.4   The principle of dynamic binding of information

A cognitive relation between a set of concepts 'A' and 'B' is realized by a 'detour' via a set of conditions 'C'. For instance, 'A' may represent the information to be processed and 'B' the set of potential consequences. The relation between both sets is then established by a set of conditions 'C', which are used to project the set of information 'A' to the sets of actions 'B'. The set of conditions 'C' represents the specific experienced-based memory about the concepts. The act of memory retrieval is then represented by the probability of the dynamic binding. The more the set of conditions in memory matches the set of information, the higher the conditional probability and the better is the set of information reflected in memory (e.g. Jones et al., 2003).

The generic architecture of memory   The dynamic binding is an indirect relation between a set of concepts A and B via C (Shastri, 1988). It is more comparable to the nature of an electric field rather than a hard-wired relationship. Figure 4.5 shows the application of the architecture to the example with the plants and animals. It represents the same facts as represented in Figure 4.3. However, the information is structured differently. The generic memory structure distinguishes two essential aspects of memory.

- A concept layer containing objects, attributes and options for actions.
- An experience layer that represents the interrelations between the concepts.

This experience layer is an essential new aspect with high importance for understanding cognitive processes. It contains all the interdependencies between the concepts a person once experienced. It represents some kind of 'binder nodes', which become active only if they receive simultaneous activation from nodes on the concept layer.

The use of this architecture in combination with the dynamic binding avoids the combinatory explosion. It can be shown that this topology is far less complicated and requires far less coding of information than any other architecture and also allows encapsulation and polymorph relations (see Sträter, 1997/2000). Encapsulation is enabled by the experience layer and polymorph relations by the concept layer. Overall, it can be seen as the principle architecture of human memory or as a representation of the internal world.

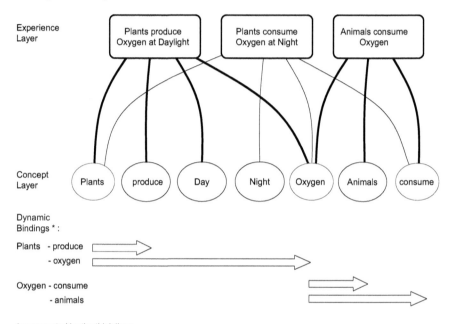

Figure 4.5   The principle architecture of human memory and the dynamic binding of information

Most network approaches view memory as a flat and unstructured set of stable states (Shastri, 1988, p. 340). Dynamic binding, however, proposes a system architecture where each concept may be represented as a neuronal sub-net. The circles in the figure above therefore represent neuronal sub-nets. In the sense of the neural network of the brain, activation spreads in a regular manner through the net and not in an unstructured massive parallel way as often discussed. Semantic retrieval of information is assured by an activation of appropriate nodes in the memory network. The activation of certain nodes represents the answer to a particular question or query to memory.

The generic memory structure is not making a particular distinction between an object, a property or a property value as classical memory models do (e.g. the goal notation models). This uncomplicated view is already suggested by Bochenski (1954, p. 45) who stated that knowledge-entities could be factual (eidetic) or procedural (operational). This view suits the law of simplicity of nature. Whether a concept is representing an object, a property or a property value is an inherent aspect of the concept itself (see Pearl, 1988; Shastri, 1988 p. 340).

It is important to realize the distinction between concepts and experiences. Experiences lead to skills but are not skills. Experiences seduce humans to recognize information from the external world as true or not true (theory of coherence of truth, Keller, 1990).

Although the generic structure appears too easy, memory cannot be much more complex than these two layers. The 100-step rule would not allow a more complicated structure of memory, supposing that each experience or concept itself consists of a neural sub-net with the depth of about 15~30 neurons. Besides the law of simplicity of nature, the architecture fulfils the important maxim that cognition is experience-based. All memory contents required for this architecture can be derived from experience. No additional assumption about the origin of information is required. In the course of the chapter it will be discussed how logical operations, changes in the level of abstraction and composition of experience can be performed. The easy architecture with the minimal number of layers also allows understanding the fast real time performance of the brain.

The use of dynamic binding in the analysis and prediction of behaviour   The distinction of memory according to concepts and experiences is powerful in order to explain the different behaviour of humans in different contexts, although a majority of contextual aspects is similar. It meets the event-based maxim of cognition and explains why behaviour is context dependent, how biases are generated based on specific experiences, and how specific dynamic situational conditions influence human behaviour. The following example shows how the structure can explain human behaviour.

In the Japanese town Tokai Mura, a criticality accident occurred on 30th of September 1999 at a uranium processing plant. It caused the death of two workers due to radiation and called for evacuation within a 10 km radius of the plant. The media reported soon after the accident that workers' unsafe actions deviated from the approved procedure and caused the accident rather than failures or malfunctions of hardware (Furuta, 2000).

In the accident, the process chain was changed by the personnel and a safety vessel, limiting the uranium that can be processed to a non-critical amount, was circumvented in order to process more uranium. The procedure had been established for a long time for low-enriched uranium but higher enriched uranium was processed that day. The increased enrichment led to an over-critical situation in one of the processing vessels and a nuclear chain reaction was initiated causing a release of radioactivity into the environment.

The regulatory body reported that the accident was a typical organizational accident. A popular explanation of the accident is that the company pursued efficiency rather than safety due to a highly competitive climate in the international nuclear fuel business (Furuta, 2000). One major causal scenario for the accident was the economic pressure due to international price competition, which motivated the plant management to pursue efficiency rather than safety (Yutaka, 2000).

Beside the many aspects in organization and management, which will be discussed later, the staff involved in the accident performed the deviated operation based on their own experience with the system.

A couple of cognitive aspects are interesting to note. The three workers wanted to finish the job quickly, because the crew leader wanted to keep his promise to give the process coordinator a sample of the product just after the noon break on 30 September and wanted to start the training of newcomers in the afternoon (Kohda et al., 2000). The working space was small and located far away from the main building, which made communication difficult.

The workers had a positive experience with the changed procedure based on former events. The substance never became critical because they had been processing 15% enriched uranium for a long time. The difference in operation was only the fact that 18% enriched uranium had to be processed that day. This slight increase in the portion of radioactive material caused an over criticality of the substance and a chain reaction, which then led to the release of radioactivity.

Kohda et al. (2000) stated that the action performed by the operators was not a deviation from their intended action and no error occurred in their execution. The issue was their experience within the job. They were used to processing low-enriched uranium, which was usually processed in the plant, and they were not aware that they were dealing with high-enriched uranium that day. However, the leader of the crew may have recognized the difference because he asked a supervisor whether the usual procedure would be all right for the higher enriched uranium. Kohda states further that

- The process was a usual process practised for a long time and never any problem occurred (positive experience).
- The work changed process was easier to handle and less time-consuming.
- The illegal procedure was confirmed by the supervisor.
- There was no change in the enrichment level of uranium to be processed for about three years and the treatment of high-enriched uranium was unusual.

Figure 4.6 gives a rough picture of how the concepts and experience of the operators were probably represented on the working level. It gives a good explanation, of why the workers acted the way they did. As the example shows, the distinction of memory into concepts, experiences and consequences provides an expedient basis for the explanation of a couple of cognitive effects in dynamic situational conditions. Regarding the Tokai Mura event these are:

- Explanation of why the action was performed based on the given experience.
- The lack of awareness of negative consequences, because negative experiences were not made (or maybe made in other contexts like training or education).
- The learning from the positive experience and establishing of strong but wrong rules because of the positive experiences made with less enriched uranium.
- Explanation of attitudes and biases as a result of the experiences made.

- The immediate utility of doing the action and the increased cognitive effort involved in thinking deeper about the risk of acting like this or in complaining to management.
- The lack of safety awareness and, as a consequence, risk-taking behaviour because negative consequences were not experienced, a utility regarding efficiency was given and the cognitive effort could remain low.

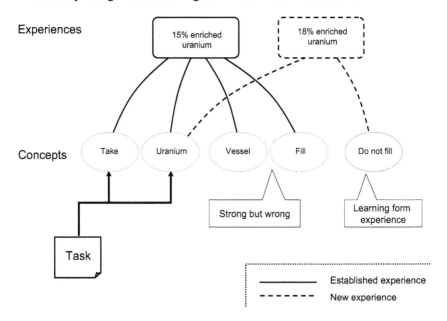

Figure 4.6   The experiences of the staff in the Tokai Mura accident

The Mathematical Calculus of Reasoning

Mathematical modelling of cognitive binding Any cognitive performance must be describable in a mathematical sense. The question of computational effectiveness lies in the synthesis of explicit inference, like in logic reasoning, and massive parallelism, like complex signal detection, in a very short time period. Both features were discussed mathematically as encapsulation of entities and polymorph relations, which are realized by the generic architecture of memory distinguishing between the concept layer and the experience layer.

The architecture of concepts and experiences, together with dynamic binding, also allows an easy and efficient description of the mathematical basis for a cognitive act of human behaviour. Beside the memory structure, the basic cognitive acts would require an institution where the central cognitive acts of coupling, binding and levelling take place. This institution does not have the purpose to process semantic information but to judge whether all the variation from the external world can be controlled by the internal world. The semantic meaning of the information is processed in the memory structure. The institution may be called the *Central Comparator*. As discussed above, the central comparator is most probably located in the gyrus cinguli region. This conclusion does not vote for a comeback of the central processor hypothesis of cognition but for complementing parallel and techno-morphologic approaches into a harmonized and balanced model of cognition. The example of the plants in Figure 4.7 explains the functioning of the central comparator with the concept and experience layer.

The processing of a central cognitive act requires a layer to launch the query and to retrieve the results (Kohonen, 1988). Whether a fact of the external world is represented in the internal world can then be simply represented by the condition of activation equality: if all information disseminated into the memory structure comes entirely back to the requesting layer, the representation is given. If not all activation is fed back, some aspects of the external world are not represented in the memory structure. The layer requesting and retrieving the activation from the memory structure is therefore a central comparator, which compares whether the information fed back at a time t_2 fits to the information sent at a time t_1.

In the example of the figure, the query is the question 'What produces oxygen?'. Leaving the dashed lines and circles aside for a moment, the propagation of this fact activates 'oxygen' and 'produce' with 50% of the activation each. The experienced relation 'plants produce oxygen' will be activated and merge again all activation sent by the query layer. A backward dissemination of the activation-level achieved activates the concept 'plants' by 100% and the concept 'animals' only by 1/2=50%. Therefore, the only concept receiving all disseminated activation is 'plants' and this is the correct answer.

The calculus changes slightly if the dashed-lined part of the figure is taken into account. In this case, still 1/1 of the activation will reach the experience 'plants produce oxygen' and 1/2 will reach the experience 'plants consume oxygen at night'. However, after back-propagation of the activation achieved on the experience layer, the concept 'plant' will only achieve 66% of the activation sent out. The statement 'Plants produce oxygen' is not entirely true. The amount of activation fed back to the query layer, however is more than 50% and can be considered as mostly true (in this example). This result can then be transformed into a dichotomous statement by using a threshold function. Using the maximal available 66% as a threshold in a succeeding cognitive act would make the statement true. This lowering is equal to changing the level of abstraction (further details on the calculus can be obtained in Sträter, 1997/2000).

The comparison of the sent activation with the fed back activation describes the levelling of the central cognitive act. The calculus is simple enough to be performed with one cycle of the cognitive act and therefore explains herewith how complex experience-based behaviour is possible even within a reaction time of about 200 ms. It also shows how sensitive the reasoning of the cognition is if new experiences are generated. A further important aspect of the calculus is that the result of a first iteration can be used for further elaboration of a statement in succeeding iterations.

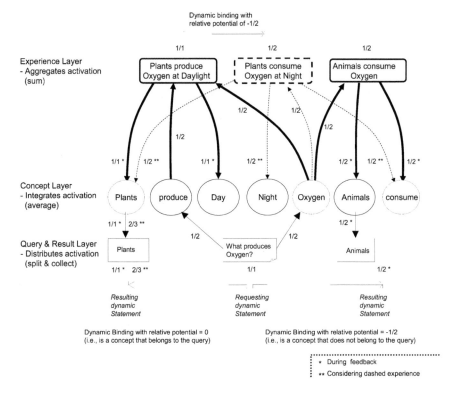

Figure 4.7   The calculus of information processing

The use of the calculus for cognitive levelling and iterative reasoning The function of the query layer is to establish the cognitive coupling by disseminating the activity into the internal world represented in the memory structure (the cortex area). The layer also needs to judge whether all activity coming from the external world can be mapped to an accordant internal energy level and to initiate consequences in the case of mismatch. Possible consequences after a single iteration are:

- The activation level a concept achieves matches the disseminated activation (i.e., concept belongs to the query).
- The activation level is smaller than the disseminated activation (i.e., concept does not belong to the query).

Resolving mechanisms In the case of perfect match, the cognitive levelling does not have to intervene. The statement is regarded as true. An action can immediately be issued. In the case of mismatch, it remains internal and the central comparator needs to resolve it before being able to issue an action. The central comparator seems not to have the ability to ignore information, which can be concluded from the fact that the P300 and N400 components cannot be avoidable even if consciously intended. In psychological terms, a stable relation in the current activation of the central comparator represents the goal of the person wants to achieve. A stable relation means that there is no mismatching information, which might lead to inhibition or negative feedback. This goal is superior to new information either coming from the external or the internal world. Unstable relations in the current activation of the central comparator may lead to the following resolving mechanisms after one cycle of a cognitive act was accomplished.

- Change the threshold level so that a back-propagated activation makes a concept true, even if it has not achieved 100% of the sent information. If, in the example above (including the experience represented with the dashed lines), only 66% of the activation is achieved by the concept 'plant' the comparator might lower the threshold to 66% to make the statement 'plants produce oxygen' true. Mathematically speaking no additional calculation of the new threshold is required. The difference in the activation can directly be used to lower the threshold in such a way that the result becomes true. This lowering of activation is equivalent to the change in the level of abstraction. The lower the threshold the higher is the level of abstraction in the cognitive act. The lowering of the threshold extends the number of concepts revealed as true in a central cognitive act. This feature lowering of the threshold of the central comparator changes the level of abstraction in reasoning and allows for widening the intersection between the concepts regarded as true. This allows building classes of concepts, higher-levels of abstraction, or finding commonalities between concepts. In turn, the higher the threshold the more precise is the reasoning. No particular inference machine is needed as in classical techno-morphologic approaches.
- Inhibit the activation of those concepts and experiences that prevent reaching 100%. This can be achieved by issuing emotional control actions. Emotional control actions may launch certain transmitter substances. The substances inhibit the spreading of activation into those experiences having a negative dynamic relation and which 'disturb the intact picture of the world' (e.g. inhibit 'consume' with providing an activation of -1/2 to the concept). Emotional reactions may also amplify those experiences which 'fit into the picture' by reinforcing the spread of activation (e.g. activate produce with a value of 1/1 in order to compensate the influence of the contradicting experience). In case certain information is inhibited by the central comparator in order to achieve a fit, even more activation from the external world is required to bring this information back into the processing loop.
- Refine the query by taking into account additional information from the external world or by refining the concepts involved in the query. In the example above, this might lead to the result that either the concept 'day' is taken into the query to make the statement true. The refined query would be 'What produces oxygen at day?'

The central comparator and processing loop allows modelling reflexive reasoning and the threshold for changing the level of abstraction meets the fact that humans find solutions even in cases where external information is independent from the internal knowledge base. These aspects are essential criteria for a cognitive model (Shastri & Ajjanagadde, 1990). In terms of Anderson (1996, p. 246-250) the change of the level of abstraction is the main mechanisms by resolving difference between the internal and external worlds.

Link of cognitive performance to behaviour Understanding human behaviour error would require knowing the memory structure of the person (concepts and experiences). In safety assessments, where trained persons are assessed, concepts are most of the time reflected and assured by the fact that the staff are trained and familiar with the working environment. Usually underrepresented in safety assessments are the experiences the persons have with the concepts. They explain rule breaking, routine or even deliberate violations, which cannot be understood by using a classical techno-morphologic approach. Due to the law of uncertainty, 'classical' approaches have to limit themselves to the level of concepts and cannot explain how they are interrelated.

The generic memory structure and the central cognitive acts of coupling, binding and levelling result into a couple of psychological aspects explaining how human behaviour and error develop. The specific combination of central cognitive acts and the generic memory architecture explains the psychological aspects relevant to understanding this human behaviour in dynamic situational conditions. This is outlined in Figure 4.8, which summarizes the discussion on the mathematical calculus. An action j is performed only if the central comparator has achieved a consistent alignment of the external world and the internal world. The action has not necessarily to be compliant with the task i. Action j may be a compliant action to task i, or to a violation. No coherent or consistent relationship between sensory information and actions can be postulated, although this would be desired for the sake of easy safety assessments (Hofstätter, 1973, p. 149 in Katzenberger, 1967, p. 42).

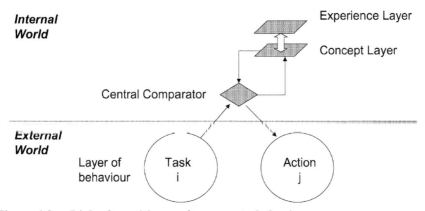

Figure 4.8   Link of cognitive performance to behaviour

Elements Determining Decision-Making

Higher-level cognitive properties From the mathematical perspective of cognitive binding, the generic structure of memory, and the generic nature of the central cognitive act provide the building blocks for higher-level cognitive properties, which are of importance to understand human behaviour in any more complex environment.

The architecture does not depend on rule- or memory-processors (as a classical expert system would need). It suggests retrieval of memory contents with activation and inhibition based on simple calculations. The memory structure allows considering the role of context in human cognitive processing. The process of reasoning is fast and parallel. Generating and retrieving a request is achieved by a distribution of activity, which relies on an easy calculation. General impacts from global neurotransmitters are related to the cognitive system via the mechanisms of the central comparator. The difference between divided and focused attention is given by different thresholds of the central comparator (the lower the threshold the more is attention divided).

Motivational and emotional aspects are linked to the role of the central comparator with its close link to the endocrine system. In the case of mismatch, positive as well as negative stress may result from the endocrine system attached to the central comparator. Biases, attitudes, and heurisms are to be understood as established experiences within the memory structure, which affects the use of the concepts. Attitudes may be seen as experiences that are established long term. Biases can be seen as rather short-term experiences and are related to declarative knowledge. Heurisms are established experiences on procedural knowledge.

Distributed nature of decision-making The cognitive control loop does not assume a specific stage for decision-making and hence no specific processing stage can be made responsible for an error in decision-making. Moreover, this section will show that decisions are a natural consequence of the generic memory structure and the cognitive processing loop. Consequently, the terms 'decision error' or 'violation' become insupportable because both are a result of the experiences made by a human being.

In the case of mismatch, each comparison also contains implicit, unconscious aspects (those where external information and internal information fit to each other) and explicit, conscious aspects (those where external information and internal information do not fit to each other and another iteration of a cognitive act is required). In the discussion of the ladder-model and the GEMS-model, common properties for errors in skill-based, rule-based and knowledge-based behaviour were identified in information-driven aspects, the alignment of information and goals and the process of generating decisions. It can be concluded from the approach of dynamic binding that any error in decision involves three main cognitive aspects:

- Information-driven aspects: The information represented in the central comparison before the cognitive loop starts (coupling).
- Alignment of information and goals: The information processed through the network of concepts and experiences and the formation of a proposition for an appropriate solution as described by the calculus of reasoning (binding).
- Process of generating decisions: The activation sent and compared with the received activation from the internal world, whereas a mismatch represents a conflict between given information and internal representation (levelling).

Looking at the interrelations of these three aspects, human cognition is basically object-driven. As the basis of their cognition, humans always need something that is processed in a cognitive act. Speaking in philosophical terms, the famous law of cognition by Descartes (Cogito ergo sum) always needs some object (cogitatum) to establish this law. To analyse cognitive errors, the functional relations between these three aspects have to be investigated.

Cognitive hysteresis Such a view on cognition implies that cognitive behaviour and cognitive errors are a result of an iterative process between perceived situational conditions and cognitive mechanisms or, according to Neisser (1976), between the external and the internal world. In every situation, operators are on a certain 'path of diagnosis' that either may be optimal or erroneous. If the path is erroneous, a certain amount of 'cognitive effort' is necessary to change to the optimal path. This cognitive effort depends on the *Situation Awareness* of the operator (Meister & Hogg, 1995). However, indications for changing the path usually do not lead to a change in behaviour without indolence. The goals of the operators play an important role since they lead to a sluggish reaction on situational cues (e.g. operators want to avoid being the plant shut down in order to save a loss of millions of dollars although safety-rules that suggest shutting it down).

According to Norman (1986), these sluggish interrelations lead to a *Cognitive Hysteresis*. Hysteresis means: if a goal or an information is considered once in a cognitive act, it needs more counter-arguments to shift the act into a different direction than would be required if the cognitive act cycle is still open for new information or goals. Norman described the situation and effects that were observed due to cognitive hysteresis as

- similarity of actual and perceived event,
- partial and incomplete explanation,
- search for confirmation.

Partial explanation means that contrary information is inhibited in order to achieve a stable relation in the central comparison. A search for confirmation means that people do focus on information and goals that confirm a once established stable relation, neglecting disconfirming information. Once a decision is taken, we usually do not revise it, even if counter arguments come up, because we would have to question our own internal world. Such a *Cognitive Hysteresis* is a result of the generic memory structure of concepts and experiences. A parable of physics can be used to get an idea of how a hysteresis develops. Figure 4.9 explains this general effect of hysteresis using the example of magnetic fields.

Each magnet in the figure may represent a certain experience made. If the contradicting information comes into the context of the internal world, which has the same amount of existing information as in the internal world, the activation in the internal world can be changed easily (a). In case the internal representation is already settled (b), the same new information is insufficient to change the internal representation because the internal picture supports each other and pulls the intended wish of change back into the over-all picture ('-' repels '-' and '+' attracts '-'). In the example of the plants, the new information of day versus night does not change the general statement that plants produce oxygen. However, it is much more difficult to accept new information if the old picture was never questioned over the years. As an example, the car industry is currently very reluctant to introduce cars with side-sticks instead of steering wheels. Although technically possible, the acceptance of the users with decades of experiences with steering wheels is expected to be too low.

Information contradictory to a current hypothesis needs more 'energy' to persuade the current loop to change direction (or to change the current mind set). The entire processing loop is aligned to the current flow of activation and this makes the change of the representation difficult. Depending on the current activation within the cognitive loop, the same information may change the mind set in one situation and not in another. The time the information enters the loop may change whether it is considered or not in the decision-making process. As the metaphor of magnetic fields also illustrates, sufficient external information may suddenly change the mind-set. If the external information reaches a level of more than about 50% of the energy within the actual mind-set, the entire mind-set suddenly shifts completely as we know from persons involved in accidents or other traumatic experiences.

Hysteresis has two main effects that can be observed in human behaviour: first, one needs more energy to change the mind set once an experience or concept is established as a coherent picture in memory. This can be easily seen with increasing age and experience. New information hardly changes the way of working, the more experienced the staff on a certain position (e.g. Rothaug, 2003). In accordance with the resistance to change is the perception of time. The less new external information that gets into the memory structure, the faster is the perception of time. Second, it shows that, once the change is accomplished (after high energetic efforts), the opposite of the former mind-set is often taken (e.g. former smokers are often the most aggressive non-smoker opinions; the trust in a relation is more diminished after one escapade than if escapades are the rule). Trust in new technologies is also a good example. The more an operator trusted a new piece of equipment and it failed, the less he can be persuaded to continue using it. Trust is difficult to gain but easy to loose (Bonini, 2004; Kelly et al., 2003).

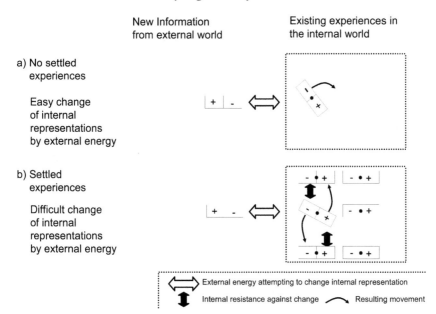

Figure 4.9   The role of settled experiences and generating of mind-sets

Hysteresis also shows that training is not necessarily an activity that increases safety. If the training content is not well proven to serve safety, the routine gained by the training may lead operators to ignore important aspects of a situation. Prominent accidents due to over-training or wrong training are the nuclear incident at Three Mile Island and the aircraft accident of a British Midland aircraft in 1989:

- During the incident at Three Mile Island (TMI), an operator switched off the high-pressure injection pumps because he was trained to combine a high water level in the reactor pressure vessel with the potential to over-feed the vessel, which risks the vessel breaking. Switching off the high-pressure injection was the trained required action in such situations. In this particular case, steam bubbles pushed the level high and, in fact, no cooling of the core was provided after the switch-off of the injection pumps.
- The other example is the aircraft accident of a British Midland plane in 1989 where, in a situation where one engine failed, the pilot was trained to combine a certain acoustic symptom with a malfunction of an engine. Because the noise came from the opposite side to that of the failed engine, he switched off the working engine. This resulted in the loss of all engines and a loss of the aircraft (Airsafe, 2002).

Cognitive tendencies The resolving mechanisms within the central comparator are required in the case of any irregularity between the internal and external worlds. The result if irregularities are disturbances of the cognitive control loop can potentially result in decision errors. In terms of the ladder-model, decision errors are a result of the mechanism to change the cognitive level. A couple of authors investigated the role of decision-making in complex environments (overview in Mosneron-Dupin et al., 1997, Gerdes, 1997; Evans 1989; Thelwell, 1994).

Thelwell (1994) investigated the triggers for the resolving mechanisms and named these as complexity-shaping factors. In this study, complexity is understood as a result of a number of properties. They include the root cause of a disturbance (symptom masking; required diagnosis effort), the spread of information (location of controls; memory load), confusion (misleading cues, faulty data) extent of information collection effort (awareness; knowledge), obviousness (saliency of alarms), attentional demands (amount of information and distractions), and severity of the disturbance as well as time pressure and task density.

Mosneron-Dupin et al. (1997) investigated the behavioural outcome of the resolving mechanism by investigating accidents and incidents in nuclear environments. They found evidence of resolving mechanisms in actual behaviour of operators and called them cognitive tendencies. Additionally to the four generic cognitive tendencies related to the processing loop, Theis (2002) could identify a fifth tendency in an experimental investigation on car driving behaviour. The fifth tendency is characterized by actively investigating the situation and enriching the picture of the external world.

The cognitive tendencies observed during incidents show the bounded rationality of human beings (Kahneman & Tversky, 1979). Although we are capable of performing many things in parallel, humans are uni-dimensional information processors. They need certain stability in their internal world to perform. This can be achieved either by fixating on an established solution, by ignoring information, or by simplifying the goals. However, the literature has also identified modes where humans add new information, or imagine new goals. In interviews with witnesses of incidents, it can be observed that they add information to their testimonies, which they could not have perceived from the external world but added from their internal world. Witnesses said, for instance, that they have seen how two cars crashed, but a deeper investigation showed that this was impossible from their position in the landscape. They added visual information based on what they heard at their position (e.g. Semmer, 1994). In addition, Dörner (1997) showed in various experimental settings that people add new goals or attitudes to their strategy in problem solving, if this fits a better overall picture of the results they have achieved. Based on extensive research on errors in planning and decision-making, Dörner & Schaub (1994) distinguished these tendencies into problems of available information and characteristics of the problem-nature. Vester (2002) identified six errors in dealing with complexity from the information processing perspective. The findings support the following cognitive tendencies:

1. Fixation on established coupling: The desire for consistency of the internal world is the driving factor for cognition (Festinger, 1957). Any cognitive act leads finally to a fixation on the decision drawn during a cognitive act.
2. Information ignorance (utility-oriented): We cannot process all our knowledge (internal world) and all the information that the external world provides us in every second of our lives. A well-known human characteristic (e.g. from

political decision-making) is the strategy to ignore information available either in the external world or in the internal world if this would lead to internal inconsistencies in a given situation. In such cases, we give preference to our internal goals and start to perform in a utility-oriented mode.

3. Expectancy-driven (habitual): Much more difficult than ignoring information is to 'cheat' or change the internal world. It is difficult to change experiences, which manifest into skills, habits, goals, attitudes or intentions. Once goals are abandoned this usually leads to polarization and more negative statements about the former position since this is necessary to achieve internal consistency of own experiences and to reduce dissonance.

4. Information and goal elaboration (rationale): In cases where high demands in cognitive coupling coincide with dissonance, the operator tends to be reluctant to act until he has found an appropriate solution for the mismatching constellation between information and goals. The human acts rather rationally but the cognitive processing loop does not find a proper exit.

5. Adding information from the external world (enrichment): Instead of ignoring information or goal setting, an additional resolving mechanism is the adding of information in order to achieve a stable relation. This enrichment of the picture with information from the external world requires a proper feedback of system states and procedures well aligned to the spectrum of tasks. In case these prerequisites are given, no internal resolving mechanism is needed (no information ignorant or expectancy-driven processing).

The effect of cognitive dissonance (Festinger, 1957) is reflected by these basic resolving mechanisms. Dissonance may relate to emotional effects in order to improve the levelling in the central cognitive act. The memory structure distinguishing between concepts and experiences additionally induces certain sluggishness into the overall process of the cognitive processing loop that leads to hysteresis effects between the information considered and the goals followed. Figure 4.10 relates the resolving mechanisms to the cognitive processing loop.

It could be shown that these tendencies can be observed in a relatively stable manner in a range of industrial settings. Besides nuclear and car driving, they could be identified in aviation (cf. Linsenmaier & Sträter, 2000) and occupational health (cf. Bubb, 2001). The approach of cognitive tendencies suggests combining systematically situational conditions and cognitive aspects by considering coping-strategies an operator may choose. This approach therefore may be considered as a way of combining the usual concept of processing stages (like perception, decision and action) and processing phases (skill-, rule-, knowledge-based). It has been successfully applied in Human Reliability Assessment showing that a cognitive error can be understood as an inadequacy of the central cognitive act regarding one or several of these aspects (OECD, 2000). The following discussion shows why human behaviour is prone to errors in dynamic situational conditions. The figure also contains the other determinants of decision-making. These are the memory structure with the distinction of concepts and experiences, which are mainly related to the effects of hysteresis, and the relation of the central comparator to emotional aspects.

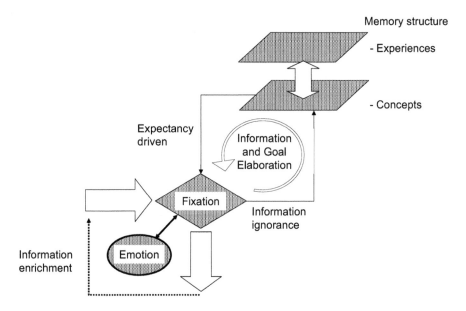

**Figure 4.10   Elements determining decision-making related to the processing
            loop**

Fixation on established coupling   A person may be fixated on a certain task to
accomplish. In such cases, the difference between external available information
and the internal representation of the world concerning the aspect of interest is not
recognized (fixation). A slight mismatch is then not considered as critical (selective
or focused attention). For example, driving the car to the office by a usual route,
we may fail to notice slight differences of the external world, like changes of
traffic signs – an effect called stereotype fixation (Reason, 1990). One solves a
problem by focusing on certain dimensions of the problem (cf. Dougherty, 1992;
Dörner, 1997).

    As a result, people tend to search for verification instead of falsification once
the stable relation is established (Reason, 1990). Fixation also contains an element
of sluggishness (cognitive lockup). People hold on to their initial hypothesis, even
if falsifying evidence comes up. They tend to focus on initial faults in disturbances
neglecting other information and tasks while concentrating only on the information
that is related to their prevailing hypothesis (Reason, 1990). In complex working
environments, it can be shown that fixation leads to a reduced questioning attitude.
Humans tend to try to solve problems by themselves, before asking others to help
them (IAEA, 2001a).

    As fixation is mainly allocated at the central comparison, it is strongly related
to emotional aspects of behaviour. Tunnel vision may be a result of a resolving

strategy by reinforcing the cognitive control loop using emotional triggers in order to achieve a stable internal world.

Information ignorance/reduction A mismatch may be perceived but not all of the relevant information is collected (bounded or uncompleted search). In such cases, the external available information implies more opportunities than the operator(s) currently want(s) to cope with. If we have, for instance, a certain hypothesis about the reason why a technical system failed, we complement information supporting the hypothesis and tend to ignore information that is apparently present but contradicts this hypothesis (confirmation bias) and prefer to work out the existing problems (eagerness to act). Eagerness to act is a means to reduce the cognitive strain as fast as possible.

For example, if during driving by car to the office the usual way is blocked, we usually take a known alternate route without diagnosing whether it is better to stay in the jam (because it is a short one). Information that is more familiar is preferred to be kept in the cognitive loop (familiarity bias) and information easily imaginable is given too much weight (capture effects). The imaginability is also related to the memory structure of concepts and experiences (Tversky & Kahneman (1974). Sangals et al. (2002) investigated the neural correlate for this tendency in EEG experiments and found that expectations are reflected in the force people use in performing actions. He found in a choice reaction task under time pressure that people tend to put more force on the buttons under high time pressure compared to situations without time pressure.

Goal reduction A mismatch may be perceived but the hypotheses are not elaborated well enough in the diagnosis process (well known as frequency gambling, bounded rationality, and routine reaction). In such cases, the internal representation of the world concerning the aspect of interest is providing more opportunities then are currently relevant given the external available information. For example, car-light failure is more likely assumed as a broken lamp, not as a fuse-failure.

During the incident in the Davis-Besse nuclear power plant, for instance, a prescribed safety-procedure, which would have made the plant inoperable in the future, was ignored and a different recovery action was initiated, which had the chance of circumventing the undesired impact on future operation of the plant (Reer et al., 1999). In many other incidents it can be seen that this conflict between efficiency and safety usually leads to the ignorance or circumvention of safety rules while giving preference to economy instead. Simon (1955) explained this simplification of complex problems with the limited capacity of human cognition. Parallel alternatives are not adequately compared (Payne, 1980, in Wickens, 1992).

Bounded rationality is often assumed as being due to the limited capacity in memory. However, it seems rather to be an inherent property of the resolving mechanism of the central comparator.

Dissonance of information and goal within the processing loop The information from the external world and the internal world may mismatch. Mismatch produces dissonance and may be a result of inappropriate information collected or inappropriate hypotheses generated from the internal representation (i.e., a mixture of information reduction and expectancy-driven processing). For example, the car is broken and one has to decide between the two alternatives that have pros and cons (buying a new one, repair the old one). As a result, one may jump from one alternative to another, treating each of them very superficially (Reason, 1990; Wickens, 1992). We become more and more confident the better the information fits our internal representation.

Because it is easier to find a stable solution in the processing loop on specific sub-elements in such dissonant situations, small details may become of high importance because they lead to stable relations and can be solved. More important issues may be disregarded not because of unawareness of their importance, but to reduce the cognitive load of not being able to find a solution, respectively stable relation (Dörner, 1997). One tends to settle for satisfactory rather than optimal solutions, which is sometimes called central bias (the avoidance of extreme values).

Another way out of the dilemma of not being able to find a solution is to attribute global causes instead of the combination of causes. Causes for errors might be 'the regulator', 'the management' or 'the technology'. General attributions of errors or responsibilities are classical examples of not being able to find a stable relation between external information and internal representation. In the final consequence, the threshold for reasoning is adapted. Believing in God is the highest possible level of abstraction one can take to find a stable relation and to explain the relation of external and internal worlds.

From the psychological observation of emotional reactions during mismatch situations, it needs to be concluded that emotional aspects are issued together with dissonance between internal and external information. This can also be shown to be the effect of moods on the cognitive tendencies. In depressive moods, the probability of failure is estimated to be higher (Gerdes, 1997).

Adding information from the external world One strategy to resolve dissonance in uncertain situations is to start seeking for additional information in order to find a stable relation. The selection and use of information depends on the ease of availability of information (availability bias; Tversky & Kahneman, 1974). The search for additional information will be postponed if the expected utility is not achievable (Simon, 1955; Reason, 1990). The utility relates to the effort required to make external information fit the memory structure and the activation of concepts and experiences. If certain concepts or experiences were already used in the reasoning, they are already primed by the activation they received. Related to priming is the sampling bias and recency effect of memory. We consequently allocate more attention to recent information or information available in the short term (Wickens, 1992). On the other hand, people make insufficient use of searching for key information if this is not directly available (Reason, 1990). They fail to seek information for all alternatives and limit the information search to the most available alternative, irrespective of its appropriateness (availability heurism).

Memory structure Many properties of human cognitive processing stem from the architecture of memory distinguishing between concepts and experiences. The relation between both is bi-directional. This can be concluded from the inference bias. Humans often assume that the probability that A will cause B is the same as the probability the B was caused by A (cf. Wickens, 1992). Humans are no probabilistic reasoning machines. Tversky & Kahneman (1974) showed that humans tend to judge more on the basis of perceived similarities rather than statistical calculations.

Recently gained experiences increase the estimation of the probability of the event. For instance, the probability of an aircraft crash is estimated as higher immediately after an accident and estimated as low if nothing has happened for a long time. This effect of hindsight bias is also essential in expert judgement methods on risk assessment, as they are often used in Human Reliability Assessments. The distinction of concepts and experiences is also reflected in the Halo-effect (Katzenberger, 1967). If one has to judge about one particular aspect of a person (e.g. his ability for mathematical calculation or language), there is a tendency to judge the ability based on other aspects of the person (like the clothes he wears, appearance etc). The interrelations learned by experiences play a significant role for judgements (King & Kitchener, 1994).

The strength of an experience depends on the probability of its confirmation by prior experiences (Katzenberger, 1967 p. 41). Experiences drive future behaviour. The higher the number of experiences the more the relationship between concepts that is represented by the experiences is considered as being true. Successfully used alternatives are preferred to other solutions, even when not adequate for the current situation. Tversky & Kahneman (1974) named this effect 'anchoring'. Experiences may lead to illusory validity. If a causal relationship between concepts is expected based on prior experiences the validity of this relationship may be assumed to be true for any case (Chapman & Chapman, 1969).

Our memory architecture makes cognition very efficient by representing the most relevant experiences and makes even complex information available even in fractions of a second. However, the architecture was built for a less complex environment (i.e., for living in a jungle where the overall system did not change that much during humans' lifetime) and reveals pitfalls in highly complex, rapidly changing working environments. The experience-based architecture is the reason for the strong influence of habits in our life, for stereotype takeover of problem solutions regardless of their applicability, and to so-called strong but wrong actions without considering whether the action is safety critical or not. It also leads to the overconfidence in the experience we made and to the deficiency to understand the experiences made by others.

As the use of experiences in reasoning is dependent on the number of experiences made, the influence of new experiences may become less with each experience made. New information is increasingly judged as of less importance.

Mental Utility and Comfort

Cognitive effort The research on risk taking behaviour shows that the dissonance is resolved either by expectancy-driven processing or by information reduction. For instance, Tversky & Kahneman (1974) identified the tendency to prefer a certain but low gain, to an uncertain but high gain, which means that that information is considered which gives the most utility within the given situation. Long-term effects play a minor role (ballistic action; Dörner, 1997). The same holds for long-term losses. People tend to prefer certain costs to uncertain and high costs (Wickens, 1992). The latter tendency is the core business of any insurance company.

The cognitive processing loop is a positive processing loop. That means that the memory structure and the comparator are operating without any negative feedback in case the external world and the internal world fit each other perfectly. This characteristic is some kind of passive information processing with minimal energy consumption. Inhibition would require active processing, which is attached to defence. In turn, the positive feedback characteristic leads to carelessness if nothing serious happens for a long time (complacency error). Tversky & Kahneman (1974) observed this effect already and named it the 'level effect': the decision depends on whether the same problem is formulated in terms of profits or in terms of losses. In addition, Wickens (1984) manifested that it is far easier for humans to process positive statements than negative statements. Double negations cause more thinking effort. Krifka (in GIHRE, 2004) could show that well-performing crews use more encouraging, positive words.

Another hint for the preference of the cognitive apparatus to operate in a positive feedback loop is the so-called Zeiganik-effect. An increased memory activity can be measured if tasks are not accomplished or stories are not finished properly. Rather than inhibiting an unfinished story or an open outcome, humans prefer to continue with open stories to find a stable relation. 'Happy ends' are therefore a must for the nature of the cognitive processing loop.

In addition, the art of rhetoric reflects this nature of the cognitive processing loop: sayings such as 'Do not think of a rainy weekend' need first a generation of the picture of a rainy weekend before they can be inhibited. Another example is the expression 'Not that I disagree with your opinion', which articulates disagreement with the opinion. Negative feedback in general leaves the person in an unstable relation and causes cognitive dissonance and emotional affection to deal with it. The ambiguous, unstable situation introduced by negative feedback causes fears and worries, because the human has to leave the area he is familiar with (Dörner & Schaub, 1994). Consequently, it is easy to understand that positive feedback is seen as a key success strategy for education and management.

Overall, humans minimize the cognitive effort allocated to a task in order to keep spare capacity available for unexpected events. Therefore, immediate adverse consequences are considered more important than delayed, more grave consequences (e.g. global warming). If consequences with a major negative impact are the expected outcomes of human intervention, people tend to not perform any action or delegate the decisions to others.

The individual mental utility makes the internal world superior to the external (called 'self-centred bias' or 'egocentric biases'), which leads to the fact that humans ignore statements contradicting the own motives or affecting the self.

After major accidents for instance, people are not able to talk in detail about the event and their role in the causation because they are aware that any public statement might have serious consequences in terms of potential legal interventions. After accidents in occupational health, people start to bend the real event in order to protect themselves. Intervention techniques were developed to deal with the cognitive effort after an accident or incident (Mitchell & Rahmann, 2004).

Mental utility The resolving mechanisms of heurisms, biases and attitudes and emotions try to reduce cognitive effort. Motivational and emotional aspects are linked to the role of the central comparator with its close link to the endocrine system. In the case of mismatch, positive as well as negative stress may result from the endocrine system attached to the central comparator.

Any means of reducing cognitive dissonance is of value for the internal equilibrium. Trimpop (1994) observed that the estimation of the likelihood of an event depends on the value of the outcome. Mental utility is an immediate utility for achieving the equilibrium state. In many incidents it can be seen that the conflict between efficiency and safety usually leads to the ignorance or circumvention of safety rules while giving preference to the economy instead (Sträter, 1997/2000). Violation and rule breaking therefore has to be understood as mental effort caused by an imbalance between internal representation and the external world and not as an unwillingness of the person who violates the rule. Violation and rule breaking are a solution of a complex problem rather than a deliberate act of harming the system. Vester (2002) identified the following stages to deal with complex systems, which also describe how violations develop:

- One tries to survive by intervening with seemingly useful interventions (simple symptom-measure relations). First, violations occur and one receives feedback (or not) on the acceptance of the initial rule breaking. These usually lead to problems, because the interrelations in the system are neglected.
- One gets interested in the interrelations of the variables and their dependencies. Situations where rules need to be followed are distinguished from those where rules can be broken.
- One starts to investigate the interrelations of the variables, tries to gain deeper understanding and tries to understand what went wrong with the initially started actions. Attempts of systematic violations occur.
- One starts to change or to violate the system structure. Violations become routine and general practice.

Mental utility develops with the resolving mechanisms. While first seemingly useful interventions may still be rational, habits develop resulting in efficient, utility-oriented exploitation later on. At the final stage, mental utility always goes hand in hand with fixation on the established experiences. The practice becomes a 'common right' for the user going beyond the rational descriptions states in procedures.

This process is not a deliberate one. Violations are a natural outcome of experiences gathered and transformed into the memory structure, and ill-defined arrangements in the working environment, such as impracticable procedures.

As the example of violations shows, mental utility considers short-term effects and quick wins. It disregards side-effects and long-term effects while striving for salient effects (Dörner & Schaub, 1994). A considerable mental effort seems to lie in the shift of goals. As soon as humans have to give up goals they try to preserve their feeling of their own competence and become dogmatic in their thinking, which finally may lead to dogmatic entrenchment on opinions that are becoming unaligned to the reality. Humans then develop the tendency to keep going even if there is an indication that the own hypothesis might be wrong. This behaviour is coupled with overestimating the likelihood of success if the outcome is desired (Fraser et al., 1992). We know the influence of dogmatic entrenchment from historical and recent influences of religion or political attitudes on behaviour.

Dogmatic entrenchment can be observed in any working environment. Related to the levels of operations introduced at the beginning of this book, it becomes a problem in the tension between regulation and operation. For instance, safety cases in the nuclear industry developed into an exercise for regulation and show little reflection of the real safety issues (Kirwan & Sträter, 2002; Sträter & Zander, 1998). A typical outcome of conflict situations is that safety problems are not openly discussed.

The desire of mental utility therefore may lead to critical situations if utility ends up in the situation of dogmatic entrenchment. Conflict management approaches are known means to overcome such situations (Doppler & Lauterburg, 2002). Vester (2002) defined six problems of complexity, which can be interpreted as a means to prevent proactively the negative effects of mental utility. The principles encountered by him are observable in virtually any working environment, like product development (e.g. software), regulation of an industry, operation of a technical system, or even daily problems:

1. Thorough planning and explicit description of the goals in order to prevent the need to perform repair actions.
2. Analysis of the dependencies in situations in order to understand feedback loops and interactions between ideally all parameters of the system.
3. Define halting points in order to anticipate the problem that actions may be biased and may lead to non-reversible system states.
4. Consider side-effects, which may exist, in order to prevent effects not envisaged and specified in the system but leading to unwanted system states.
5. Do not over-steer the system (drastic changes) in order to be able to repair or perform remedial actions.
6. Avoid authoritarian management or control, as this would increase the conflict between the person in control (or seemingly in control) and others (satisfied

utility of one person may be mutually exclusive to the mental utility of other persons).

Group utility So far, individual cognitive processing was discussed as a result of the architecture of the processing loop. In addition, the effects of group performance and group perception rely heavily on the individual architecture of cognition. The immediate link is given by the false consensus bias. Individuals think that their internal world is also the internal world of others. If I am convinced about certain aspects of the external world due to my experiences, I assume other persons have this opinion as well (see Gerdes, 1997).

Groups develop shared values, which are shared experience in the terms of the cognitive processing loop. Group conformity develops. Generating common experiences is therefore a central aspect for building groups. The type of experiences is of less importance to develop the shared view.

As the individual mental utility makes the internal world superior to the external one, group thinking makes the group world superior to the outside-group world. The risk of being wrong is estimated to be lower when it affects others. Increased risk taking due to shared responsibility may be the result and humans tend to be overconfident in their own abilities (sometimes called 'illusion of control'; Wickens, 1992).

This happened in Chernobyl. The crew performing the incorrect action and circumventing the reactor protection system was the most effective crew, and was honoured for its effectiveness shortly before the accident. Voluntary controllable or familiar risks were perceived as low.

The resolving mechanisms discussed for the individual usually also hold for a group. Group conformity is a stable relation within a group with common experiences and common values. In the case of non-conformity, dissonance comes up in the group and the same effects that are used to align the internal world to the external world are projected now on the group (e.g. Schuler, 1995):

1. Fixation on established values and group norms (fixated).
2. Ignorance of new information coming into the group (utility-oriented).
3. Preference to salient shared group values (habitual).
4. Disputes in the group on the way forward (rationale).
5. Consultation of different opinions to enrich the picture (information adding).

Cases 1 and 2 represent a more dictatory organization of groups, putting pressure on members of the group who do not agree with the general idea of the group. Cases 4 and 5 represent rather democratic organizations. Case 3 represents a decision making process of groups. Katzenberger (1967) investigated the conditions for the low validity of decisions in social environments and found social and cultural experiences as superior drives for stereotypes. Such social or cultural biases influence the decision for personal and professional development.

Cognitive comfort The stability of the internal world is more important than that of the external world. Mental utility and group utility are used to achieve a comfort for the cognitive system. Comfort was intensively investigated in anthropometric design (e.g. Bubb, 1999) but it is also of high importance for the cognitive system. For instance, the BFU report on the mid-air collision at Lake Constance stated, according to the fact that there was only one controller on the working position although two were required:

> Officially, this procedure [to assume only one controller in the night shift] did not exist, but had been in practice at ACC [Air Traffic Control Centre] Zurich for many years. This arrangement made the night shifts for the controllers more comfortable. ... Even though it was an unofficial procedure, it was known to and tolerated by the management. (BFU, 2004, p. 75)

The situation was more comfortable for the group of controllers and for the management as well. Gillard (1979) made an investigation on comfort and defined five factors for anthropometric comfort. Relating this research to the concept of cognitive comfort, one may infer the following factors to achieve cognitive comfort in a working environment:

- Low dissonance between internal and external world (higher dissonance accepted in foreign or unknown environments).
- Good reflection of the expectations and goals of the person or group of persons within the situation.
- No permanent disturbances (even if they are only small, permanent disturbances and adaptations are causing discomfort).
- Smooth transitions between different system states in dynamic evolvements.
- Move forward by making new experiences from the person's or group's perspective.

Comfort includes that humans have an inherent motivation to build up new experiences. However, if the change is too slow or too high for the specific cognitive processing loop of an individual, this may cause cognitive discomfort.

Cognitive discomfort may increase up to a point where humans start avoidance behaviour in order to protect themselves. Humans start to search for external reasons for their failure, rather than blaming themselves and place their own desires as superior to situational factors (Mosneron-Dupin et al., 1997). This defence leads to underestimating the likelihood of undesired outcomes for oneself (Kontogiannis & Lucas, 1990). Self-reflection and self-criticism on the actions done or the goals set may decrease.

Attitudes The use of attitudes is effective in order to anticipate future events and to prepare the organism for future efforts (Reichart, 2000). Roth (1967) allocated attitudes as long-term outcomes of specific task demands (Figure 4.11). Biases, attitudes, and heurisms are to be understood as the result of established experiences within the memory structure, which influences the use of the concepts. Attitudes may be seen as experiences, which have been already established for a long term.

Biases can be seen as rather short-term experiences and are related to declarative knowledge. Heurisms are established experiences on procedural knowledge.

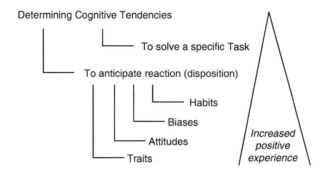

Figure 4.11   From tendencies to solve tasks to habits, attitudes and traits

The Perspective of Cybernetics and Information Theory

The preceding section described how decision-making theories and characteristics can be explained by the principal cognitive processing loop. The loop has a second distinct advantage of also enabling the inclusion of reaction times, memory span (sometimes called short-term memory), logical reasoning and learning. Cognitive load (mental workload) can finally be discussed as a result of these cybernetic considerations. The key for the relationship between these psychological issues and the processing loop is the architecture of the central comparator.

The Architecture of the Central Comparator

The event-oriented topology of the memory structure explained the basic properties of human reasoning. The next question would be how higher logical functions and mental workload can be understood as a result of the same cognitive system. This question is related to the way the internal world is accessed by the central comparator. The central comparator has the task of judging whether the activation fed back from the memory is equal to the activation sent out. This comparison needs to consist of two elements:

* A mechanism of how the level of activation is sent out, how inhibition is managed, and how a decision is made by lowering the activation threshold.
* A distribution of the activation into the right areas of the memory structure.

Managing the level of activation by coupling, binding and levelling The resolving mechanisms, as discussed in the previous section, are mainly a result of arousal management. Arousal management comprises adjustment of the threshold and determining the amount of inhibition for a central comparison act. Arousal management is related to consciousness.

The processing of cognitive performance is often distinguished into conscious processing and unconscious processing (cf. Schneider & Shiffrin, 1977). The nature of consciousness has been investigated from Freud (1936) to Wickens & Hollands (2000). Usually it is distinguished between:

- Unconscious processing: In the techno-morphologic models like the ladder-model, unconscious processing is understood as simple S-R (signal-response) behaviour. However, unconscious processing is more than a simple S-R reaction without any other cognitive process happening in between. It comprises automated cognitive behaviour like association or identification of well-known objects and therefore uses memory as well. Unconscious processing is uncontrolled, open loop cognition of information from the external world. Therefore, unconscious processing is to be understood as a positive feedback cycle of the processing loop.
- Conscious processing: Conscious processing means active control within a situation. Any higher operation, like active search or the generation of new hypotheses based on given information, requires conscious processing. The human cognitive system acts in a closed loop with the external world in order to resolve contradictions between the internal and external worlds. Recursive reflections (like logical operations, learning) on the internal world (like own thoughts) also require closed loop operation of the cognitive processing loop. Therefore, conscious processing is to be understood as a negative feedback cycle of the processing loop.

The distinction between conscious processing and unconscious processing should not be mixed up with the skill-, rule- and knowledge-based level of the ladder-model. Conscious processing makes the cognitive act clearer but does not change it (Pauli & Arnold, 1967, p. 134, in Katzenberger, 1967, p. 24). Unconscious and conscious elements cannot be separated by the level of skill that is to be used by an operator to solve a problem, as is sometimes suggested (i.e., skill-based is purely unconscious and rule- and knowledge-based fully unconscious). Knowledge-based behaviour, for instance, is always accompanied by elements that are processed unconsciously. Consciousness is rather the criteria to move from one level of behaviour to another. The unconscious aspect is an essential aspect of the errors on that level, because errors on the knowledge-based level may occur due to inadequately applied resolving mechanisms. Any of the resolving mechanisms (e.g. reduction of the information) finally establishes a positive feedback loop. As a consequence, the reliability of human cognition is always a result of experiences made, concepts learned and the situational aspects not conforming to the concept or experiences.

Understanding unconscious and conscious as either positive or negative feedback allows us to explain the concept of 'consciousness' as a result of the central cognitive act, consisting of dynamic binding, the concept layer and the experience layer (cf. also Piaget, 1947). Negative feedback in general leads to a stabilization of a status (Dörner, 1997) and hence negative feedback within the cognitive processing loop leads to a stable internal world.

In the case of a perfect match of the activation, a single pass in the network is sufficient for inference. A fast reaction in about 200 ms is possible. The network acts in an unconscious, positive feedback loop. The speed of the act is independent of the content of the cognitive act. It may by a simple reaction on a signal coming from the external world, as well as the intuition or bright idea on a solution for a complex problem.

Conscious, as well as unconscious, cognitive processing is always a result of the given information, the internal world and the binding of both. Due to the fact of positive feedback, humans are in principle reactive in judging about the negative consequences of their behaviour, if the negative consequences are not already part of their experiences. Awareness about negative consequences would require an inhibition (negative feedback) as it is caused by a mismatch situation. An inherent characteristic of our positive feedback algorithm is hence that we have to make negative experiences to learn.

In the case of mismatch, the schematic actions are not sufficient any more and a human has to 'break out' of the uncontrolled positive feedback loop and turn into a controlled mode (Mechsner, 1998, p. 68). The next cycle is introduced in the processing loop by threshold adaptation, inhibition of activation, or changing the activation by considering new external information. Any of these resolving mechanisms require a next iteration of a central cognitive act. Due to the change of activation, the activity network acts in a negative feedback loop. The issuing of negative feedback is equivalent to performing a conscious act. The negative feedback loop may continue several times, which is equivalent to discursive reflection on a problem.

Consciousness can be understood as threshold-adoptions or inhibition within the cognitive processing loop. This creation of consciousness can be observed as mismatch negativity in the EEG. The psychological effect of the resolving mechanisms is known under the concept of *Cognitive Dissonance* (Festinger (1957). If there is a disturbance in human habits, which cannot be remedied by automatic behaviour (positive feedback), then this generates a cognitive dissonance and an attempt is then made to correct this dissonance by means of efforts to achieve consonance (negative feedback). This explanation is consistent with the findings of Miller (1956). If different varying signals need to be processed (i.e., iterative processing with negative feedback) then this cognition is sensitive to time pressure. However, automatic behaviour is rather robust against time pressure (i.e., positive feedback). Stable relations seem to be robust enough not to be impacted by time pressure.

There is no difference in the mechanism of mismatch and cognitive dissonance between low-level cognitive processes like perception or recognition and high level cognitive processes like decisions under uncertainty, as the models of Reason or Rasmussen would suggest. The difference can be explained by the number of iterations the loop was already processing.

The launching of a new iteration and a further cycle is equivalent to the rule-based behaviour as stated in the model of Rasmussen. The next iteration may lead to the result that all the activity fits, and the dynamic relation between sent activation and received activation is established. The relation becomes stable and an action can be launched. The higher the number of iterations to make the statement true, the more the behaviour becomes knowledge-based. This results into the psychological cognitive levels known as the ladder-model. The higher the number of iterations, the more the central cognitive act becomes a problem-solving process with inferences and reasoning. This view on the distinction of skill-, rule- and knowledge-based behaviour as being more related to the number of iterations fits to the fact that the cognitive levels are rather to be seen as a continuum than as a dichotomy (Wickens, 1984). Virtually any higher-level model can be derived from this generic functioning of the cognitive act. As an example, the TOTE (test-operate-test exit) units (Dörner, 1976) can be represented by the basic cognitive acts of mismatch, coupling, match and action. Figure 4.12 reflects the TOTE cycle within the cognitive processing loop.

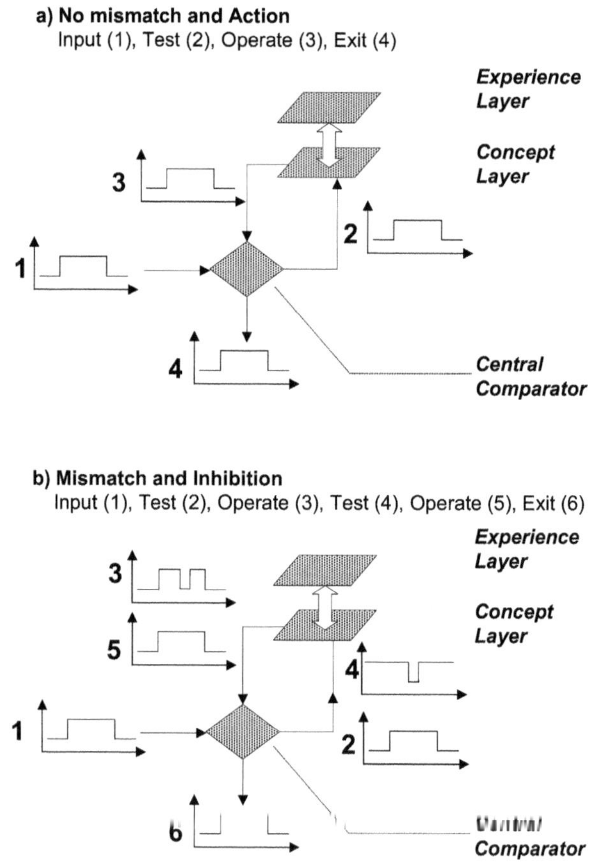

Figure 4.12   Iterations of the cognitive processing loop

In example a), all external information is represented in memory and an exact match is achieved. The cognitive processing loop achieves a solution after one cycle. In example b) a mismatch is noticed between external information (1) and internal information (3). The central comparator inhibits the mismatching activation (3) in the second cycle (4) and herewith achieves a match of external and internal information (5) that then leads to an action (6). The scientific aspects of each techno-morphologic approach can therefore be represented by this approach. This holds for the models discussed in the genesis of modelling cognition in safety and also for additional models derived in ergonomics or cognitive science.

This inhibition in cognition is an energy efficient process. The cognitive system only needs to spend cognitive resources on those aspects that bring the system out of the equilibrium. Within the approach of the processing loop, this means that any further cycle of reasoning uses the mismatch difference measured in the preceding cycle directly for the inhibition in the next cycle, without complex calculations, i.e. the deviance is already the energy that is further used in the cognitive cycle.

Distribution of activation The central comparator stores the actual alignment of the external and internal world using one threshold value. Once new information comes into the loop this information is sent into the cortex area and it is expected that the same amount of activation comes back. The central comparator just needs to notice differences, not the semantic meaning of the information (this is done by the memory structure in the cortex area). The distribution of activation needs to break this activation into mathematically resolvable pieces to perform a cognitive act.

The central comparator does hence not need to store the semantic meaning of the information to perform this task. It is only required to 'mirror' those properties of the information that allow a mapping of the information to be aligned. An organizational principle of the central comparator is to order the information-patterns in such a way that the information fed back can be clearly assigned to the stored activation. The properties need to be structured orthogonal to each other in order to prevent interference. The mathematical calculus required is known from the design of databases (Wiederhold, 1981). Generally, the information needs to be broken into independent pieces. This process is called normalizing of the data. The Hamming-distance describes the distance between the internal and external world and the required iterations to resolve the mismatch (Ameling, 1990).

The so-called Gestalt Laws provide a first hint on the properties of the central comparator, which enable orthogonal structuring. Gestalt Laws describe the identification of shapes and come long before semantic processing of information. Usually the following main Gestalt Laws are distinguished (e.g. Guski, 1989; Leeuwenberg, 1968):

- Spatial proximity: Closely arranged elements are perceived more as a unity than separated ones.

- Common movement: If two objects are moving simultaneously, they will be brought into a causal relation or perceived as identical.
- Similarity of shape: Two squares will be perceived more as a unit than a square and a circle. Similarly, framed elements will be perceived more as a unity than not-framed elements.
- Tendency to good shape: If a pattern is incomplete, it will be completed.

The general principles behind the laws are space and time (Kant, 1966), which can both be understood a result of the recursive nature of the processing loop. Gibson (1973) could validate the neural correlate of spatial representation in his optical flow theory. It is known from habituation experiments that humans are able to identify the direction and spatial location of an acoustic or visual signal long before the reason for the orientation reaction is known by the individual. Togerson (1962, p. 9) investigated the basic properties that can be distinguished by the cognitive system. These are physical length, width and colour. They come before one can identify objects like a house, a person etc. From research on motor control and learning, a third dimension to take into account is the force (Schmidt, 1988). Fitts (1964) showed that the learning of motoric actions is related to the number of iterations to experience misalignment of muscle-forces to the motor task to be accomplished. The same holds for the judgement on sensations, which finally led to the db-scale for the sensation of sound, for instance.

The basic operations of the central comparator can be summarized as more, less or equal activation, which are applied to the three basic parameters of space, time and force. This results into the following basic dimensions:

- Space: higher or lower, before or behind, left or right, close or distant.
- Time: Sooner or later.
- Force: bigger or smaller, stronger or weaker.

According to these dimensions, a semantic meaning can be applied to any external information. Objects are distinguished from each other by their relative position. An object is bigger, greater, has more colours compared with another object, and defines therefore its specific characteristics ('ars specifica' according to the Aristotelian rules of logic). The competitive economy is one outcome of this generation of objects by relative judgement as well as the individual power distance.

Figure 4.13 illustrates how letters are processed by the central comparator. The figure uses the classical example of processing the expression 'The Cat' with ambiguous information on the letters 'H' and 'A' (Selfridge, 1955). The central comparator assures that the letters 'H' or 'A' are correctly processed according to the context in the word (in this example due to spatial arrangements).

The generic properties of the central comparator are important in human error investigations and attentional interference. For instance, expert judgements are easier to be made based on relative scales than on absolute scales (e.g. Comer et al., 1984). The 'law of comparative judgement' (Thurstone, 1927) is hence a consequence of the central comparator. This property also explains the generic error classification used in all human error methods distinguishing between qualitative errors (like too much, too little) and time errors (like too early, too late).

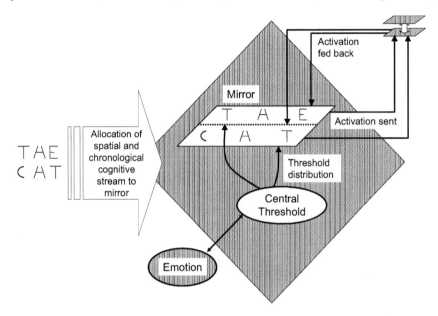

Figure 4.13   Processing of letters by the central comparator

There is a certain time of inconvenience on getting information from the external world into the processing loop. Information from outside cannot be processed appropriately because the comparison-loop is busy with other tasks. Humans start to get problems in aligning information if there is a conflict in respect to these basic dimensions. Interference occurs if the information has an overlap in one of the organizational principles of the central comparator. However, interference does not occur if the overlap is of another nature. Therefore, tasks can be easily performed in parallel even if they are mentally more complex, as long as spatially or temporally different.

As Low (2004) reports, large performance trade-offs between a cursor positioning task with other spatial tasks can be found (Wickens, 1984), while such tasks do not interfere with arithmetic tasks. Instead, mental arithmetic tasks interfere with other tasks where sequences need to be remembered. Fukuda & Voggenberger (2004) showed that operators of a nuclear power plant start to get into difficulties when talking to a communication partner if they had to manage an operational task with similar dynamics to those the communication task required. Further interference effects were extensively studied by Wickens (1984) and led to the multiple resource theory (MRT) distinguishing between verbal and spatial coding as well as auditive and visual information. The MRT was always competing with other resource theories, such as the time-sharing theory on mental resources (GIHRE, 2004). The architecture of the central comparator and cognitive processing loop integrates both approaches and integrates resource theories with higher cognitive reasoning.

Logical reasoning A tempting reason for the use of the computer metaphor was the human capability of performing logical decision-making. The information theory could validate this view as it could describe the basic relations mathematically (Attneave, 1974). Human reasoning is an explicit semantic and logic process. The semantic part of human reasoning is explained by the generic memory structure of experience and concepts. The process of reasoning is a result of the central comparator.

Any reasoning starts with a concurrent situation between the external and internal world. As discussed, this mismatch in the central comparator is a decision problem, which can be solved either by a resolving mechanism or by launching a succeeding cognitive processing loop of the cognitive cycle. Launching the second loop is the first step of logical reasoning, i.e. finding a contradiction to the current representation of the internal world.

However, reasoning has another element, which is implication, i.e. the assignment of new information to the current content of the cognitive cycle. Implication is the act that makes us aware about our internal experiences, our internal world and finally ourselves (Mechsner, 1998 p. 64). Implication requires that one piece of information dominates another one and hence requires a comparison of both, which has to be done by a central comparator.

Implication is a conscious act of inference, i.e., an act containing negative feedback on our internal world. The most effective proof of the ability to infer one's own consciousness is the recognition of oneself in a mirror (Mechsner, 1998 p. 66). The recognition of oneself in a mirror is an inference from one's internal world as being existent. Continuous self-reflection leads to recursive reasoning and new experiences of our internal world.

Reasoning may be based on the experiences or based on concepts (Bochenski, 1954, p. 133). Experience-based reasoning is usually called inductive reasoning (or bottom up). It is using a set of specific experiences to draw general conclusions out of them, which then become concepts. Concept-based reasoning is usually called deductive reasoning (or top down). Experiences are generated based on reconfiguring concepts in a new constellation. Bottom up and top down processes are therefore not different directions in a hierarchical network but rather iterations on different types of memory (experiences or concepts).

The central comparator explains how logical acts are possible. A logical act is a particular relation of the actual activation of the mirror and the activation fed back from the cortex. Reasoning works with the positive feedback loop of the central processor. The possible relations between both are shown in Table 4.2.

Table 4.2   Logic reasoning as the relation between activation in the mirror and activation fed back to the central comparator

Relation between activation A in the mirror at time t_1 and activation fed back from memory M at time t_2					
Possible combinations: $A(t_1)$ **0 0 1 1** $M(t_2)$ **0 1 0 1**					**Explanation of the activity on the central comparator (logic expression) and Examples for resulting cognitive behaviour**
Possible relations:					
R_0	0	0	0	0	No processing (null-Function) No cognitive cycle is launched (hypothetical)
R_1	0	0	0	1	Further processing of aspects being in both mirror and memory. (and-Function) Stable relation is established
R_2	0	0	1	0	Further processing of aspects that are not in memory (M inhibits A) A sequence of external information processed
R_3	0	0	1	1	Further processing of aspects, which are in the mirror, new information from memory is inhibited. (Identity with A) Further elaboration of the activation in the mirror in a succeeding cycle like further elaboration of the situation
R_4	0	1	0	0	The new aspects coming from memory are processed (A inhibits M) A sequence of thoughts is built up
R_5	0	1	0	1	Further processing of aspects which are in the memory; information in the mirror is inhibited (Identity with M) Further elaboration of the activation in the memory in a succeeding cycle like formulation of intentions
R_6	0	1	1	0	Further processing of aspects which are not fitting to each other (Anti-valence, exclusive OR) Concentrating on the differences between mirror and memory
R_7	0	1	1	1	Further processing of all aspects activated (or-Function) Elaboration of all aspects in mirror and memory
R_8	1	0	0	0	Further processing of aspects not currently evaluated (not-or-Function) Elaboration of new aspects
R_9	1	0	0	1	Further processing of aspects equivalent in mirror and memory (equivalence) Confirmation of match between mirror and memory
R_{10}	1	0	1	0	Further processing of aspects not in memory (non-Function) Curious search for new information
R_{11}	1	0	1	1	Creative logical act of inference from activation of mirror to memory (A implies M) Implication can be used to build up hierarchical relations or to generate new contents in memory.
R_{12}	1	1	0	0	Further processing of aspects not in mirror (non-Function) Deeper thought about a memory content
R_{13}	1	1	0	1	Creative logical act of inference from activation of memory to mirror (M implies A) Creative logical act generating new hypotheses about a situation
R_{14}	1	1	1	0	Further processing of aspects not present in both (and-not-Function) Elaborating the differences between mirror and memory
R_{15}	1	1	1	1	All information is processed (one-Function) The threshold is reduced so much that any information is processed

A '1' in the table means an activation pattern is present in a certain part of the mirror at time t_1 or after another cognitive cycle at time t_2. A '0' means no activation is present. The activation in the mirror can stem either from external information or from a preceding cognitive act. The logic relations R are a result of how the central comparator deals with the activation in the further process loop. The table does only describe the hard logic, although the discussion of the cognitive tendencies showed that the logic can also become fuzzy by inhibitory interventions of the central comparator (emotional interventions). Therefore, the table does not show all the facets in each of the logic relations.

Reaction Time

Reaction time of the cognitive processing loop   The iterative nature of the processing loop provides, in conjunction with the architecture of memory and central comparator, an easy understanding of the reaction times of human to external information. Summarizing the discussion on the cognitive processing loop, memory architecture, cognitive mirror and levelling, one can assume seven main steps in a cognitive act. These seven cognitive acts break down the three central cognitive acts discussed at the beginning of this Part of the book and include execution of an action:

Cognitive coupling
- Information entering from the external world.
- Information distribution according to the activation level.

Cognitive binding
- Information query to the cortex of whether the external information matches an internal representation.
- Evaluation of the cortex areas.

Cognitive levelling
- Feedback of the results of the evaluation to the mirror of the central comparator.
- Integration of the results to an overall result.

Action
- Action execution in case a stable relation is achieved.

If one assumes that each step takes about 25 ms, the reaction time for simple reaction tasks (e.g. like switching a button as soon as a light comes on) one entire processing loop may take about 175 ms.

Figure 4.14 outlines the reaction time of a single cognitive processing loop as a result of the principal cognitive processing loop. Note that the figure provides only the principal process but can display neither the details of the brain structures involved nor the complexity of the entire processing loop.

The reaction time represented in the figure matches the amount of time it takes for a person in a middle age range to perform such a task. It is known that older people have slower reaction times (e.g. Rothaug, 2003). Reaction time may stretch to about 210 ms for the same task. Younger people may even be able to perform the same task in about 160 ms. There is also a difference over the course of the day between reaction times.

The difference can be explained by the pacemaker of the processing loop. The pacemaker is most likely the so-called alpha-frequency that can be found in EEG investigations while persons are awake. This can be concluded from the relation of reaction time and alpha frequency during aging. The alpha-frequency ranges from about 7 to 12 Hz (e.g. Schmidt & Thews, 1987), which is equivalent to a range of about 143 ms to 83 ms per impulse. The frequency is negatively correlated with reaction time. The amount the reaction time decreased for older persons is proportional to the lowering of the alpha-frequency. While young people have a median of about 12 Hz, old people slow down to 11 Hz.

The alpha frequency slows down, the amplitude decreases and spreads to adjacent areas. The individual becomes less reactive (Shearer et al., 2004). Assuming that two impulses are used to trigger the processing loop, each of the steps mentioned above may take about 25 ms if the alpha frequency is about 8 Hz.

People have only limited attention for information that is subject to time-delayed feedback (Dörner, 1997; Reason, 1990; Wickens, 1992). Bad timing in communication, where essential information is arriving in the time the processing loop is currently evaluating the information in the cortex area, leads to the situation that the understanding of the information is either delayed (Hohlfeld et al., 2004) or even completely disregarded (Fukuda et al., 2003). If information arrives at the information entering point, the delayed information can hardly be processed by humans, because the cognitive act is in a state where the information cannot enter the current processing loop.

As Figure 4.14 indicates, one cycle (from information distribution to central comparison) takes about 100 ms. This may also explain why humans have problems with delayed feedback. If new information enters the cognitive loop later than about 100 ms after the first set of information was processed, the information is not processed as part of the first loop and is seen as new independent information (e.g. Card et al., 1983). It is known from virtual reality experiments or experiments with visual and motional feedback that humans have difficulties if the visual feedback and motional feedback differ by more than about 100 ms (Bubb & Sträter, 2005).

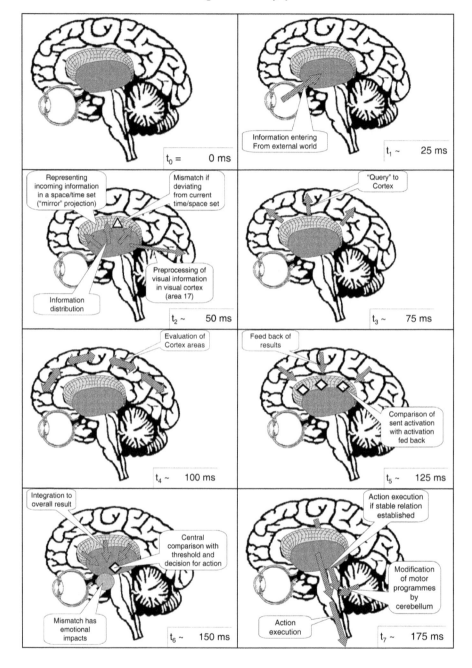

Figure 4.14   The processing loop and simple reaction time

Figure 4.15 shows this 'readiness to receive new information' relates to the central cognitive acts. The figure presents an activity diagram of the reaction time related to the cognitive acts. The figure shows a reaction time after four iterations of the cognitive processing loop.

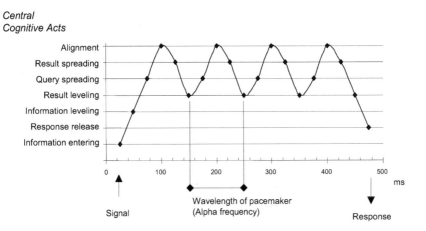

Figure 4.15   Activity diagram of four iterations of the cognitive processing loop

A good example that information cannot be processed even if very critical is the mid-air collision at Lake Constance. The controller who had to deal with the two aircraft that eventually collided also had to deal with a third aircraft. The third aircraft, on a delayed landing approach to the airport Friedrichshafen, was calling the controller literally in the second before the pilot of the Russian aircraft informed the controller about his TCAS system (Tactical Collision Avoidance System) alarm. Therefore, the information could not be processed by the controller and he gave the wrong instruction to the Russian pilot to descend.

Memory Span

Reaction time, Hick's law and memory span   It is known that reaction time increases with the number of alternatives to be chosen. However, the increase is not linear but logarithmic. Hick (1952) investigated this relationship and due to him, the logarithmic relation of the number of alternatives to reaction time is stated Hick's Law.

Hick's Law is another proof for the resolving mechanism of the central comparator. A task where a human has to choose between alternatives always implies that the alternative does not fit the current situation and needs to be inhibited. Therefore, each processing cycle has a capacity of one bit (i.e., the capacity to decide between two alternatives). Depending on the complexity of the information, the processing time per bit is in the range of 100 ms to 150 ms. The time to decide between four alternatives increases by one processing cycle. Overall, the number of information points a human can decide on in a certain time increases by 2^n-1, where n is the number of decisions. Figure 4.16 explains this relationship.

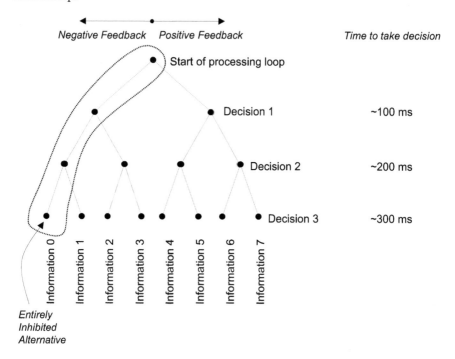

Figure 4.16   Number of decisions a human can perform per time

The number of decisions a human can perform per time is related to the memory span of information that humans can keep in a certain time-period in their minds. Miller (1956) revealed that the number of items a human can keep in memory at a time is about 7±2. Both spatial and temporal access modes follow this magical number seven rule found by Miller (1956), which indicates that this limitation is related to the central comparison and not to the distribution of information according to space and time. To memorize more than 7±2 bits, information is clustered into chunks that are accessed by a particular cue. It has often been concluded from this limitation of 7±2 bits that humans have a short-term memory module with this specific capacity (e.g. Mandl & Spada, 1988).

However, a satisfactory theory of memory span should recognize the neural structure of the long-term memory and relate short-term memory to the number of bindings per time (RWR, 1006). The capacity depends on the dynamics of the situation. Rassl (2004) showed that the capacity decreases to about three chunks during dynamic car driving tasks. Sträter (1994) showed how cues are used to access chunks in a free recall exercise on knowledge elicitation. If experts have to report about their knowledge on a certain topic, the knowledge is not extracted in a homogeneous flow but rather in 'bursts' of knowledge chunks. These packages have the size of either 3±1 or 7±2 pieces of information. There is always a pause between these packages, where the individual iterates the domain in order to find related knowledge clusters. The 7±2 pieces of information are also sub-clustered into smaller pieces of 3±1 pieces of information.

As the figure further indicates, the memory span rather reflects the depth of the decisions one can make in the time constraint of an external situation. Only static tasks can achieve the 'full' capacity of 7±2 by a depth of three cognitive cycles. The more dynamic a task becomes the lower is the depth of processing. An additional dynamic task may decrease capacity to 3±1. A highly dynamic task may even decrease the capacity to 1.

This lowering of capacity is known from critical accidents and related to the issue of tunnel vision (Wickens, 1984). The more critical a situation becomes, the more a human is not able to process all alternatives. The information a human can process is reduced to just performing one task. In addition, Dietrich (in GIHRE, 2004) showed that pilots under high workload reduce their memory span and are only able to answer yes/no type of questions. The study was based on the analysis of 16 flight crews.

In turn, this understanding also concludes that the memory span can increase to more than 7±2 in highly static, relaxing situations. Generally, it would obey Hick's Law that the memory span is 2^n-1, where n reflects the depth of processing. Depending on the depth, certain paths may be easier to follow than others may. This explains the '±2' part of the capacity law. In the figure above, for instance, information 7 could be extended another time, so that the total number of elements maintained in the memory span is nine. All 'magical' numbers used in human life can hence be derived from this mathematical relationship (like the 12 for the twelve months or apostles, as a result of 3 bits minus one branch of 4, 31 for the days in a month as a result of 5 bits minus one branch etc).

This leads to the conclusion that the 7±2 rule cannot be a rule for the capacity of a short-term memory 'device'. It was already discussed that the term short-term memory (STM) does not reflect well the nature of memory. The rule seems to be a general rule for the organization of a human information processing loop including short-term memory and long-term memory as well as the retrieval mechanism. Short-term memory cannot be seen as a store, but rather as a range of complex skilled behaviours, designed to impose transitional information on series of items that otherwise contain low transitional probabilities (Jones et al., 2003). The process of information is based upon those habits and statistical rules that direct the ease with which material is retrieved from memory (Pope et al., 2003). In other words, short-term memory is a transition probability between two sets. The higher the transition probability, the more a certain aspect will be remembered (i.e., part of the long-term memory). The lower, the more the negative feedback has to be processed to get a stable relation (i.e. short-term memory).

It is therefore suggested to return to the original term of describing the phenomena that is the *Memory Span* rather than speaking about a short-term memory (Katzenberger, 1967). Memory span is a capacity of the processing flow rather than a capacity of the static memory (Baddeley, 1982).

The timing of memory Working memory is the currently activated dynamic bound information of the generic memory architecture. Working memory hence reflects a stable relation between the internal and external worlds. The capacity of the memory span is variable according to the dynamic contextual condition a person is in (i.e., variability due to time constraints, mismatch by distractions etc).

Every cognitive cycle recalls certain information from the long-term memory (the generic memory architecture) by establishing a dynamic binding. Complex cognitive behaviour is built from a sequence of these generic cycles. The absolute duration of one cycle is difficult to determine exactly because the time the central comparator needs in cases of ambiguous matches is not known. The more stable the relation is (i.e., the more the activation received from the cortex area equals the one sent) the faster the decision of the central comparator. This was shown by Katzenberger (1967, p. 36) for association tasks. He found that associations, which only need one processing cycle, are faster if the information to be processed is more coherent. This effect leads to semantic priming, if it is related to a succeeding iteration of the processing cycle. Therefore, information that is semantically more congruent is processed faster (semantic distance of word combinations like 'doctor-nurse' vs. 'doctor-house').

Card et al. (1983, p. 42) provide a range of 75 ms to 370 ms, which is a composition of 50 ms to 200 ms for the distribution and query and 25 ms to 170 ms for the cognitive evaluation and feedback. According to Fitts (1964), the duration of the cycle may be about 150 ms. However, both investigations do not clearly describe the cognitive processing loop and still remain in a techno-morphologic model of cognition. The times also include other parts of the techno-morphologic information processing stages (like perception and action).

It may be concluded that each cycle typically takes about 100 ms, with respect to the results of EEG investigations on the P300 and N400 components as well as the alpha frequency as a general trigger. This is also confirmed by an investigation into perceiving coincidence, causality or independence (see Card et al., 1983). Typically, coincidence is perceived if two signals do not deviate more than 0 ms to 100 ms. Causality is perceived if one signal is succeeded by a second signal in about 100 ms. Independence is perceived if the two signals are delayed by more than 200 ms.

Given the capacity of the memory span of 7±2 for static tasks, the typical duration of the present may consequently be 7±2 x 100 (up to 200) ms, which is about 500 to 1800 ms. Schweigert (2003) validates this duration for car driving tasks. The duration of fixations on items relevant for performing a driving task (e.g. signals, signs, and mirror) is exactly in this range. Katzenberger (1967, p. 44, 49) found that the maximal duration of the presence is about 10 s. Given that the memory span may also increase according to the calculus 2^n-1, this is equivalent to about 7 bits.

The constraints of the memory architecture The memory span depicts the retrieval constraint of the cognitive processing loop. Memory is challenged if additional information is needed to get a stable relation (i.e., if the external picture does not fit the internal one). This causes an additional cognitive cycle. Retrieval constraints are therefore related to the experience and concept layer. In a working environment, they may be predictable on a statistical basis (e.g. x% of persons have experiences for a certain task). However, the statistical basis may differ according to education or cultural differences, for instance.

The generic memory architecture has a coding constraint due to the encapsulated and polymorph nature of the concept and experience layer. Often long-term memory is understood as being de-facto unlimited in the amount of information that can be stored. However, the event-oriented architecture of the cognitive system also has its limitations, which is due to the generic memory architecture. These are (cf. also Reason, 1990):

- Recency effects: Events that happened recently are recalled easier than events that occurred longer ago.
- Frequency effects: Events that occur frequently are recalled easier than rare events.
- Manifestation effects: New experiences are sorted into the existing set of experiences. Events that are experienced first form the basis of the memory (as habits, biases, attitudes, traits). Later events can only change this basic experience by a little (i.e., the amount of change by one experience given the number of experiences n is 1/n).
- Logarithmic perception of time: The more experiences are gained, the shorter an objectively given time period is perceived because the new experience is integrated with the experiences already made (i.e., the objectively given time period is then perceived as an integral function of all experiences).

The constraints are an essential side-effect of the economic nature of memory. The principle for retrieving and storing information is efficient, mathematically feasible and provides effective solutions most of the time. The memory architecture and dynamic binding are used to avoid the problem of combinatory explosion and represent the experiences made in a way that human beings can effectively deal with uncertainty. However, the number of experiences to make and the number of exceptions to learn are increasing with an increasingly complex technological world and therefore the algorithm reaches its limit. Human errors are the result as the system has an inherent sluggishness in dealing with too many changes after a set of experiences was learned.

Learning

The role of positive and negative feedback for learning The generic memory architecture consists of concepts and experiences. The often-stated hierarchical organization of memory seems to be more a result of the processing loop than it is a real hierarchy of the concepts or experiences. Both are rather flat clusters of neuronal nets. Hebb (1949) found that neural nets learn by the frequency of occurrence, known as the Hebb-rule. The Hebb-rule is suitable to explain the positive feedback aspect to further learning new concepts or experiences or to strengthen existing ones. The essential criterion for learning new aspects is coincidence. Learning therefore is always context sensitive learning (e.g., Keller, 1987); we recall the contextual conditions under which we learned certain aspects (e.g. remembering of the teacher when one learned mathematics while recalling a certain mathematical rule).

The experience layer explains this event-oriented nature of memory. This nature of memory may, for instance, lead to errors after retraining. Even if a controller is retrained on new equipment, he may show in a certain traffic situation his old habits, even if he had not showed it for quite some time.

On the other hand, the negative feedback and hence the mismatch in the central comparator are essential for the differentiation of concepts and experiences. To differentiate concepts or experiences, the neuronal net needs negative feedback. Learning can only be performed consciously. As an example, the consequences of actions cannot be learned in one go with the action itself. Further differentiation requires another loop. No action can be learned in full awareness of the possible consequences.

Training is often used to learn the technically required actions (skills) but less to learn the consequences of an action in different contexts. Therefore, potentially negative consequences do not become part of the memory and errors of commission or violations may be the result. Awareness about negative consequences may also deteriorate due to decay or due to further positive experiences afterwards. If positive experiences continue, the mathematical portion for the negative ones again decreases.

The progression of learning The learning of complex facts is a cumbersome process, as we all know from learning mathematics, physics or languages at school. For instance, it takes about half a year to learn all the letters and to combine them to the first set of simple words. Only afterwards, we learn to construct more complex sentences. However, the process of learning is a reduction of memory load in order to deal easier with the threats of the external world. It is similar to transforming kinetic energy into potential energy in order to have external threats better controlled (i.e. the 'kinetic energy' is a conscious process while static energy is the memory structure). Learning reduces memory load in order to avoid relearning of already established concepts, to split complex experiences into elements that fit to the concepts, to merge common aspects of experiences to common concepts (e.g. Kerninghan & Ritchie, 1988).

The progression of learning may be distinguished into content and process. The content can further be distinguished into learning of concepts and experiences. Learning of new experiences is mostly generated by the positive feedback of the processing loop and simply adding a new experience from a particular dynamic binding. However, experiences may contain more than what can be reflected in the current set of concepts. In order to explain how experiences are transferred to concepts, Piaget (1947) made the important distinction between assimilation and accommodation:

- Assimilation describes the collection of new experiences but is not changing the layer of concepts.
- Accommodation describes the process of generating new concepts from experiences.

Assimilation of new experiences is very fast while accommodation is slow because it incorporates recent experiences into the existing structure of concepts. Accommodation is some kind of consolidation of concepts. It needs reflection, thinking over, and relaxation from external pace. Many of these adaptations are therefore performed during sleep. Sleep deprivation ends up with considerable problems of the personality because the mind cannot align experiences with the concepts any more. Other hints for the generation of new concepts or the adaptation of existing ones from experiences are absent-minded situations, like staring or looking out of the window. In situations, where the human is actively involved, no accommodation can take place, because a reflection on the concepts and experiences is not possible.

New concepts or the adaptation of existing ones is based on the evaluation of the similarities of several experiences. Exceptions of rules, and contexts where certain concepts can be applied or not applied, are issues that are assimilated over the days and resolved by accommodation over night. This process of differentiation needs negative feedback loops to exclude certain parts of an experience; the organism begins to dream. Accommodation generates new concepts (or adapts existing ones) from experiences in order to prevent slow negative feedback loops next day. Overall, the accommodation also enables one to turn negative experiences into positives. The organism remembers positive events much easier than negative ones ('The good old times').

Often distinguished is learning of procedural and declarative knowledge. The generic architecture of memory does not require separating both. Concepts can be either action related or fact related. The sequence of actions is generated by the processing loop, as is done in logical reasoning.

Part III

Application

Chapter 5

Implications for Cognitive System Design

Often, simple, techno-morphologic models like the one in Figure 2.2 (in Part I of this book) are use to support cognitive design and human error assessment. Part I of this book showed a couple of aspects are underrepresented if using such approaches. Examples of accidents provided evidence that additional issues need to be represented in cognitive modelling since they have a high relevance for design and assessment. These issues were called dynamic situational factors because of their dynamic nature.

The integrated model approach of the cognitive processing loop introduced in Part II of the book provides a basis to explain and assess human behaviour under such dynamic aspects. It demonstrated an approach that is able to represent all aspects of the currently dominating techno-morphologic processing thinking but includes aspects like cognitive dissonance, control modes, or heurisms in decision-making. It showed herewith how an integrated cognitive model can solve the problems in cognitive science that are not yet sufficiently transferred into design and assessment, like the role of attention and consciousness within cognition or the interrelation of psychological and emotional issues. Additionally, Part II showed how to include the dynamic understanding of the memory span and reaction times, the essential mechanisms of human decision-making and diagnosis, or how attitudes develop from the generic memory architecture.

This chapter gives an overview for the practitioner in the field of system design of how the model can help him or her including this complex issue into design considerations. In order to demonstrate this, the chapter applies the model to actual and future problems, of including cognition into

- system design, cognitive performance (including mental workload, situational awareness, prediction of reaction times and performance);
- incident analysis and safety assessment of human contributions to risk under dynamic situational factors;
- teamwork and communication (including crew performance and cognitive aspects of safety culture).

Before going into a detailed elaboration of these aspects, the following section summarizes the essential characteristic of an integrated approach for cognitive system design, which enables us to explain the concept of mental workload, the problem of decision-based cognitive errors in automation, analysis of incidents regarding cognitive performance, and the mechanisms of crew performance based on one homogeneous approach.

Critical Dimensions of Human Information Processing

The connectionism nature of the generic memory structure does not depend on the traditional distinction between the representing memory, interpreting information and inferring action. Nor does it require a techno-morphologic resource model behind the limitations in cognitive processing. The model introduced takes the principal structure of the brain into account and does not introduce a techno-morphologic classification of human performance. For instance, the model does not suggest using an error classification, like mishearing of information, as a perceptual error, or understanding a violation as an error on the decision stage.

Rather, the model focuses on the dynamic alignment between internal and external worlds. Any behaviour is a result of how these two worlds are coupled. Herewith, the model does not blame human decision-making. Violations are understood as an interaction between human knowledge and system design. Typical preventions of violations like training therefore need to be substituted by a systemic view of cognition. On the other hand, this view on cognition requires a new view on what the essential dimensions of human performance are in order to build a classification for human performance assessment and design.

The cognitive processing loop understands higher-level cognitive aspects as being built from several cycles in the processing loop. The dimensions driving cognition are based on:

Structural dimensions of the processing loop
- Central comparator and threshold.
- Mirror of spatial and temporal order.
- Generic memory architecture of concepts and experiences.
- Emotional aspects to regulate mismatch.
- External world to trigger processing loop.

Dynamic dimensions of the processing loop
- Memory span as a function of dynamic pace.
- Reaction times.
- Processing loops.
- Learning.
- Comfort and utility.

These dimensions understand human cognitive performance as a highly parallel but uni-dimensional cognitive process. In other words, humans tend to think in causal series, not in causal nets (Dörner, 1997, Reason, 1990), although the brain consists of a highly networked structure. A stable relationship of the internal and external worlds is essential to perform an action. This desire for consistency limits cognitive human performance. This view on cognition does not make a distinction between short-term and long-term memory or between memory span and decision-making. They are different views on the same system, the cognitive processing loop. Additional aspects, like cognitive resolving mechanisms, can easily be integrated and are no more detached 'add-ons' of the modelling architecture as they were in techno-morphologic cognitive models.

The internal cognitive world (e.g. goals and attitudes of operators) is a result of activation rather than stages of processing, which explains the decisive impact of goals and attitudes on human behaviour. Classical techno-morphologic models cannot explain for instance, why goals influence what to see and what to do, as well as habits, heurisms and skills influence what to see and search for. Recursive, self-perception and its role in decision-making and trust in the system can be integrated as part of the processing loop. It does not remain unlinked to the cognitive modelling and hence unlinked to human error analysis and assessment.

Cognitive Control

The cognitive processing loop implies the need of a better understanding of high-level cognitive processes by understanding them as a result of interactions between the external world and the internal world. A couple of approaches to structure cognitive control were developed, focusing on different problems of human performance. All have in common that they allow the prediction of cognitive performance in complex dynamic situational conditions. Some of the control modes are also discussed in Reason (1990).

The contextual control model (COCOM) was developed by Hollnagel (1998) for describing human control in human reliability. Hollnagel distinguishes the control modes scrambled (occasional control), opportunistic (information-driven control), tactical (planned control) and strategic control (goal- or intention-driven control). All four control modes are equivalent to the generic control modes of the processing loop. Opportunistic control is control-driven by the external information and means launching a processing loop due to the external changes. Tactical control means that the information-driven control succeeds in a continuous stream of several control loops, while occasional control is rather related to a single, stepwise control action. Strategic control is the fixation on goals, neglecting the changes in the external world.

Jones et al. (2003) elaborated the cognitive streaming theory to describe the relation of memory and cognitive performance during communication. He could experimentally confirm that short- and long-term memory stem from the same memory architecture. He also found that tasks interfere with each other if both require sequencing elements. This interference was found between any set of tasks and apparently is independent from the nature of the task. The findings contradict the classical view of interference that would assume that at least spatial and temporal ordering is not interfering, because those orderings occupy different processing channels or resources (Wickens, 1984). Jones called this effect 'serialization'.

Voller & Low (2004) found cognitive routes as a suitable approach to predict skill changes in new automated systems in air traffic management while investigating how automation functions can be linked with Air Traffic Management training skill requirements. The decisive link between both aspects could not be achieved using the classical cognitive processing models. In contrast to this approach, they found that the controllers participating in the investigations arranged the classical cognitive process stages, like 'detect information', 'retain information', and 'switch attention' as sets of higher-level cognitive functions. They used a multi-dimensional scaling analysis to analyse the sets and found five main clusters. These clusters describe different routes of how higher level cognitive functions are generated using the generic cognitive classification. A cognitive route is defined as a sequence of cognitive processes that will be carried out together (either simultaneously or sequentially) to achieve a given function. The final set contained 24 cognitive routes. As an example, they found for anticipating cognitive behaviour, the following three routes:

- Predict, control, test and decide/select (a processing loop driven by the internal representation in memory applied to external information).
- Predict, rehearse, test and evaluate (a processing loop driven by the internal representation in memory and evaluated by external information in a succeeding loop).
- Predict, control, evaluate, image goal, and formulate action (a processing loop driven by the internal and external representation leading to iterations of the processing loop until an action is formulated).

Summarizing the approach, different control modes dynamically bind objectively and subjectively given information depending on the type of relation between both worlds. The different control mode approaches can be attached to the system ergonomics perspective of cognitive control. Theis & Sträter (2001) used the system ergonomics approach successfully to assess car drivers' behaviour using data of nuclear incident investigations. They showed that cognitive coupling allows the prediction of certain human behaviour even across different industries.

The System Ergonomic Framework for Cognitive Control Loops

Any certain technical solution (whether it is cognitively well designed or not) asks the human for a certain cognitive behaviour and for certain cognitive coupling with the technical system as long as one is concerned with task-related human cognitive performance. Consequently, the approach enabling the description, assessment and design of cognitive aspects is based on the description of the coupling of the processes in both the technical and the cognitive systems. It uses the basic assumption that any task-related cognitive performance is also somehow related to the characteristics of the technical task that a person or group of persons has to perform. The examples later on will show that this assumption is a valid one.

The properties for description, assessment and design of technical systems regarding cognitive aspects are neither fully technical oriented nor fully cognitive oriented. The properties are cybernetic or system ergonomic properties (Bubb, 1993).

Any human role in a technical system is preceded by an exchange of information between the human and the technical system. This exchange can be initiated by the mission (i.e. the set of tasks or required functions) a system designer may have prescribed to deal with a technical system, an action of the human on a technical system or the feedback the human gets from the system. All these things happen in a certain physical environment and under certain situational conditions (like time of day for instance). The same holds for a human–human relationship. Here we have the possibilities of listening and speaking and most of the time hope for a common understanding by assuming a match of our internal worlds.

As discussed, the exchange process distinguishes between internal and external worlds. Figure 5.1 shows the relation of both from the cognitive system design point of view (e.g. Neisser, 1976).

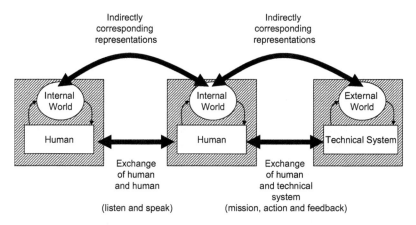

Figure 5.1   The human–human and human–system coupling

A complex working system consists of several human–human and human–machine systems. It is essential to recognize at this point that the classification of cognitive coupling provides a classification neither of the external world nor of the internal one. It provides a classification of the principal ways they can be coupled.

Based on the figure, several coupling modes can be distinguished. Table 5.1 (later) provides an overview of the coupling modes. They are described in the following and are related to the cognitive control modes proposed in CREAM (Cognitive Reliability and Error Analysis Method; Hollnagel, 1998) to some extent. Many of the distinctions mentioned here are also related to the work represented in Wickens (1992).

Isolated vs. involved processing The first important distinction is whether a human is somehow involved in the technical system or is performing in an isolated mode. In the involved mode, the internal and the external worlds are coupled with each other. Piaget (1947) assigned the term 'accommodation' for the isolated mode and distinguished this from the 'assimilation' where the people are in an actively involved mode and gain new experiences from the external world. Assimilation is also strongly related to top-down or data-driven thinking, whereas accommodation is usually top-down or goal-driven. The isolated mode is not equal to accommodation but a hint for accommodation.

The isolated mode means that the human cognition is busy with itself. No external information enters the current thoughts. Humans need such a 'timeout' for themselves; we cannot be coupled with the external world all the time. Trying to do so will result in serious psychological problems regarding the personality, as demonstrated in various researches on sleep deprivation (e.g. Schandry, 1981). The isolated mode does not imply that thinking is abandoned, but to order the different experiences made during a day or during an active involvement into a homogeneous overall picture in the internal world. Sleep, sitting while mentally absent in a meeting, or looking out of the window, are all necessary cognitive behaviours to get order into our internal world.

During examination of operational experience for instance, failures made during maintenance tasks were spontaneously recovered in the 'after work phase'. The persons had time to accommodate their experiences to their existing internal world and to 'think about the day'. They realized the error they made and informed the shift in duty. This process was observed as typically lasting about 90 min after the active involvement in the task. After 90 min off work, considerable recoveries could be observed in nuclear power plant events (Sträter & Zander, 1998).

This de-coupling of the technical system can also be observed during communication. It is a well know fact from knowledge acquisition techniques that people are not able to express their cognitive activities if they are ordering their internal world in a problem-solving situation (Sträter, 2001a).

Monitory vs. active Independent from the mode a human is currently in, the task requires the human for either an active or monitory behaviour. Both require the human to be in an involved mode of information processing (assimilation).

It is stated quite often in the literature and human factor textbooks that humans are no good monitors. This statement is often the strongest argument against automation. The processing loop and the generic memory structure suggest discussing this statement and to doubt whether it is true in this absolute sense.

Imagine a human being actively involved in the working process all the time. Certainly, he or she will make quite some errors during this permanent engagement in the working process. Therefore, we introduce breaks in the working arrangements where the people can switch from active to passive tasks. The same holds for purely passive tasks like monitoring. Monitoring the status of a display during an entire shift is as error prone as being actively engaged in the entire shift all the time. The truth is in the middle. Humans are good monitors and good performers but the balance between both has to be well established.

The reason lies in the assimilation and accommodation of memory. Highly monitive tasks with little or no change of the system mean that sooner or later a stage is reached where no new information is entering the cognitive processing loop. Experiences are fully accommodated, the cognitive coupling develops a stable relation and the loop has nothing new to process. Time stands still for the processing loop, boredom is generated and vigilance drops. On the other hand, constant changes in permanent active involvement do not provide time enough for accommodation. Experiences are building up, but not aligned with the concepts. The human operates in an assimilation mode. At a certain stage the assimilation of new experiences needs to be aligned with the internal concepts. No more additional information can be processed and time to think about the own experiences and actions is required. People start to be reluctant on new actions and need a rest.

Therefore, the general statement needs to be understood in the sense that humans are not good exclusive monitors, but human are also not reliable if permanently acting.

Closed loop vs. open loop Any task requires the human to act in a certain manner, namely controlling or initiating certain functions. While controlling requires working in a continuous loop with the technical system (e.g. regulating the flow of a valve or maintaining the altitude of a plane), the initiating of a function only requires performing a certain action. Operating is – in principle – possible without checking the outcome of the action performed (e.g. switching a pump on).

Whether an operation is perceived as closed or open loop highly depends on the time frame of the processing loop between human and machine (or human and human in communication). Typically, the range where humans see a closed coupling of their actions with the technical system ranges from about 200 ms to 2000 ms as shown in software ergonomics by Card et al. (1983). From several investigations regarding driver behaviour, it was also observed that a driver typically shifts between areas of interests in about 200 to 600 ms while scanning the scenery (Schweigert et al., 2001). If the driver is isolated for longer than about 2000 ms from the scenery, his acting seems be de-coupled from the environment.

In Schuler (1995), investigations of communication processes showed that 'good communication' requires that a closed loop between the communication partners is established, typically by short interruptions of the speaker by the receiver using words like 'OK', 'Aha' at intervals of about 2000ms. If such feedback is missing or slightly delayed (to about 2600 ms), misunderstandings or even bad communication outcomes were observed.

This typical time range of about 200 ms to 2000 ms for closed loop performance is related to the memory span (as discussed in Part II). Miller (1956) found that memory is typically limited to 7±2 information units (chunks). In highly dynamic environments, like car driving, this memory span is reduced to about 3 chunks (Sträter, 2001a) while the time frame is reduced to about 200 to 600 ms.

In time frames of 200 ms and less, this span is even reduced to 1 chunk (e.g. the effect of tunnel vision in emergency situations as observed in aircraft accidents or nuclear incidents). This reduction of the memory span is also often observed in attentional research paradigms like dual task investigations.

As a consequence, the memory span rather reflects a continuum than a dichotomy or a fixed capacity (e.g. Wickens, 1992). It can be shown that this continuum appears to be a non-linear stepwise functional that obeys the rules of information theory (see Part II of this book). The distinction between short-term memory (STM) and long-term memory (LTM) is therefore an artificial rather than a real correlate (cf. Neumann, 1992; Strube, 1990).

One- to multi-dimensional One property of the external world, with importance for internal representations, is the dimensionality of the task a human has to cope with. Investigations of operational events showed that humans usually tend to reduce their internal world to uni-dimensional concepts since there is a clear understanding about the outcome of a control action in this case.

In the case of multi-dimensional characteristics of a technical system, the human needs to imagine the interrelationships of the various dimensions. If operators in a nuclear power plant for instance have to control temperature and pressure in a vessel, this is a two-dimensional task. The operators have to imagine how these two parameters are related to each other and how they may interfere with each other during start-up of a plant.

No matter whether a technical system has uni-dimensional or multi-dimensional properties, the human only can perceive what the instruments of the technical system provide. Any interface that does not provide support to gain a good internal representation about the interrelations of several dimensions of the technical system complicates this process of imagining these interrelations. If, for instance, temperature and pressure are presented independently from each other (e.g. by two analogue instruments) and can only be controlled by two independent controls (e.g. two switches), this cognitive demanding task is far more complicated for the operator than a two-dimensional display or two-dimensional control element. In an airplane for instance, the control of position (turn and bank) and the attitude to the horizon (pitch) are controlled by one single input device, the yoke. It would be a completely inappropriate design to have two independent control elements for the horizontal position and slope.

The complexity of a technical system is therefore related to the number of dimensions the technical system has. This more functional complexity has to be distinguished from the complexity of the interface, i.e. whether these dimensions are presented simultaneously or sequentially.

Simultaneous vs. sequential processing   Simultaneous processing requires maintaining all the information necessary for the simultaneous tasks in memory (e.g. maintaining position and speed on a road requires information on the road path as well as weather and road conditions). The simultaneous processing of all information needs some co-ordination in the cognitive processing in order to perform all tasks correctly.

Sequential tasks require performance of the action in a prescribed order in order to perform the task successfully (e.g., while driving a car with manual transmission, the irreversible order is to de-clutch, change gear and release the clutch). In this case, the cognitive processing system has to recall certain rules or schemas stored in memory to follow correctly the procedure, predefined by the layout of the technical system. Simultaneous and sequential actions can be at either a conscious or an unconscious level.

Compensatory vs. pursuit Alarm systems usually provide information about a system that has become mismatched on some predefined parameter (e.g. 'pressure in vessel XY too high'). Alarms usually lead to compensatory tasks and require the human to identify the meaning of the information based on the knowledge about how high the pressure in the vessel is allowed to be. Gauge displays showing a pressure in a vessel XY require that the operator recognizes the difference as a 'more than normal' deviation and are called pursuit tasks (e.g. to detect the difference between 100 psi and 101 psi as more that allowable).

Compatibility vs. incompatibility Several elements of the external world (e.g. a switch for a pump and a switch for a generator in a power plant) can be presented to the human either in a compatible way or in an incompatible way. External compatibility refers to this similarity within the external world. It requires matching the information provided by the technical system against each other (e.g. displays and controls).

A certain element of the external world can be either compatible or incompatible to the internal representation a human has about its functioning. Internal compatibility refers to this similarity between a certain external element and its internal representation (e.g. a switch for a pump and the mental model the operator has about its functioning). Internal compatibility always implies aligning the incoming information with one's own understanding about the way the technical system works (see also McCormick & Sanders, 1983).

Compatibility is one of the golden rules in cognitive task design (e.g. in control-room upgrades, the controls in the new control room should be compatible to the ones used in the old control-room). However, cognitive design does not mean striving for compatibility all the time. Information should be presented in an incompatible way, if the devices are used for different technical functions.

A Framework for Describing Mental Load and Mental Complexity

The discussion on cognitive control leads to distinguishing more or less cognitively demanding coupling, depending on the effort needed to resolve the mismatch between internal and external. Mental load increases if the need to resolve the coupling increases. Mental utility was discussed as a major mechanism to minimize the load by achieving immediate, short-term utility.

Table 5.1 summarizes the system-ergonomic aspects describing the different types of cognitive coupling of human and technical systems or human and human respectively. The table serves as a framework for describing human complexity. Each dimension also puts a different load on the cognitive system. As an example, the higher the number of variables to manage, the higher is the cognitive demand to deal with the increased dimensionality (Theis, 2002). Arrows indicate a coupling direction. Dark arrows indicate a trigger. White arrows indicate that this path is necessary but not synchronized with the trigger. The sign (+) indicates cognitively high demanding, the sign (-) cognitively low demanding aspects. Besides the type of cognitive coupling, additional load may stem from the number of processing loops to be performed until a stable relation is achieved. The additional load stemming from this depends on:

Informational load

- The information given in the external world can over-load if too many new aspects are provided, compared to what is represented in memory. It can under-load if no new aspects are contained.

Memory load

- The experiences and concepts contained in the memory can be richer than external information, and no memory load is given, or they can be fewer, which will cause memory load and the requirement to build up new experiences and concepts.

Attentional load

- The central comparator needs to inhibit or reinforce information in the resolving mechanisms chosen to deal with a situation. One might, for instance, either constantly ignore information from outside or be curious in getting new information into the processing loop.

Complexity is seen as a highly important variable for the implementation of the single sky in Europe (the harmonization of the European airspace; Skyway, 2004). Air traffic management subdivides complexity of airspace often according to parameter, such as traffic density, traffic directions, conflict position, phase of flight, height of conflict, airspace structure, speed variation, type of conflict, weather, traffic mix (variation of aircraft types) callsign density or similar (Schäfer et al., 2001).

Such parameters are purely driven by the external part of the cognitive processing loop. The coupling and the internal representations in memory, as well as the characteristics of the central comparator, are widely neglected by such a parameter. The cognitive processing loop would suggest understanding complexity as a cognitive concept, which has a drawback to aspects like the number of sequential tasks to coordinate or the dimensionality of traffic flow.

The Role of the Processing Loop in Communication

The connectionism nature of the generic memory structure and the processing loop do not need the traditional distinctions to describe load and complexity and do not require a techno-morphologic resource model to explain the limitations in cognitive processing. The framework was proven suitable for classifying the cognitive load in communication tasks (Fukuda & Voggenberger, 2004; GIHRE, 2004). The approach of the cognitive processing loop is here a valid for both the coupling between a human and a technical system and between a human and another human.

Table 5.1 Cognitive coupling of human and technical system

Type	Visualization	Description
Type of Involvement (+)		Involved: human is in assimilation state (gaining new information) either by perceiving new aspects of the external world or by testing out internal aspects as hypotheses → interact, perceive
(−)		Isolated: human is in accommodation state (housekeeping of internal world) → reflect
Type of Task (+)		Monitory: the human is monitoring the process by collecting information and deciding about the process state (perceiving the feedback from a technical system) → monitor
(−)		Active: the human is involved in process control by collecting and combining information and process interacting; 1:1 relation between information and goal, top down processing → perform
Type of Control (+)		Closed: the task is changing over time and depends on timely demands (e.g. tracking task) → track
(−)		Open: the task is independent from timely demands → operate
Number of Dimensions (+)		Multi-dimensional: the technical system is characterized by more than one parameter that has to be brought into relation to another; more than two goals, top down processing → imagine
(−)		One-dimensional: the technical system is characterized by one parameter; one goal, top down processing → expect
Necessary Operation (+)		Simultaneously: a simultaneous operation of several controls is required; parallel processing required → co-ordinate
(−)		Sequential: the control has to be performed in a defined sequence; serial processing required → follow
Type of Presentation (+)		Compensatory-task: the system provides information for on how large the difference of task and actual state is → identify
(−)		Pursuit-task: the system provides information on the difference of task and actual state → recognize
Primary Compatibility (+)		Internal: compatibility of the mental model of a person with external information (e.g. learned, stereotype behaviour); compatibility of goals and information → align, associate
(−)		External: compatibility of different external information (e.g. displays and controls); compatibility of different information → compare

Communication and Cognitive Control Modes

Team behaviour is an important factor in recovery from system failures once an automated system fails, i.e. if it reaches a state that is beyond its designed and specified functions (e.g. a conflict resolution assistance-tool is missing a certain type of conflict in air traffic management). Those 'beyond design-bases situations' are particularly demanding and are heavily reliant on team performance.

Crew performance and communication strongly related to cognition. This relationship of communication and cognition is quite obvious since any communication needs some prior cognitive processing. The link between cognitive aspects of communication behaviour has been shown in much psychological research (e.g. Eberleh et al., 1989). Verbalization of knowledge using the thinking aloud method, for instance, shows similar effects to the ones observable in operational experience. Communication gets more difficult the higher the demands of the situation and the more difficult it is for the operator to cope. This relation of cognitive activities and communication is shown in Figure 5.2, which is based on the evaluation of operational events in nuclear power plants (Sträter, 2002). Communication is one aspect of plant safety that has an observable influence in about 10% of all human failure events.

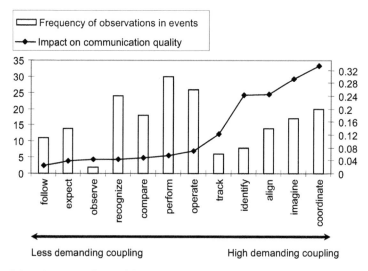

Figure 5.2   Impact of cognitive activities on communication quality

The figure shows the relative impact of cognitive coupling on communication quality. The cognitive verbs are related to Table 5.1 (above). The figure shows that the cognitive activities are a distinctive characteristic for communication problems. It shows, first, the absolute number of events where a certain cognitive activity led to a communication failure (bars) and, second, the relative importance of the cognitive activity (line). The frequency provides the number of events where the control modes played a role. The relative importance is defined by the number of events where the cognitive aspect played a role divided by the number of events where any cognitive activity was observed at all. The relative importance is hence the essential indicator to get a feeling on the impact of cognition on communication.

The impact on communication increases in the different cognitive activities from 'follow' to 'coordinate'. In situations where the operators are receptive to information from outside their internal world (i.e., in low demand coupling), communication problems are compared much less to situations where the operators are busy in applying their own thinking to a situation (in high demand coupling). The more items that are to be considered during operation, the more difficult it is to get information from outside into the cognitive loop (the cognitive mill). In cognitive states where operators are faced with several tasks and several cognitive aspects that have to be matched (i.e., during coordinating), communication problems are greater. Further elaboration of the cognitive demands using eye-tracking analysis can be found in Fukuda et al. (2003) and GIHRE (2004).

The importance of communication increases in different cognitive activities from 'follow' to 'coordinate'. The different cognitive activities can be distinguished into the following groups and are to be understood as follows (see Sträter & Bubb, 1998):

- Follow and monitor: The processing loop is driven by activities of perceiving external information (topographic cognitive behaviour according to Rasmussen, 1986).
- Expect, recognize and compare: The processing loop is driven by internal information (symptomatic cognitive behaviour according to Rasmussen, 1986).
- Perform and control: The processing loop performs more open loop cycles with active inputs of operators into the system.
- Track: The processing loop performs a permanent active engagement by inputs of operators into the system and control (closed loop).
- Identify and associate: The processing loop needs several iterations to combine external information with internal representations.
- Imagine and coordinate: The processing loop needs frequent internal iterations (run internal representations in the mind).

The analysis of the cognitive activities during communication shows that humans are able to communicate during technical task performance in open loop situations where the operators are usually also receptive to information from outside their internal world. Communication problems are much less compared with closed loop situations where the operators are busy with applying their own thinking on a situation. The more items are to be considered during closed loop operation, the more difficult it is to get information from outside into the cognitive loop. In cognitive states where operators are faced with several tasks and several cognitive aspects that have to be matched (i.e., in associative states), communication problems are superior.

Again, the human mind performs resolving strategies to cope with both the demands of communication and of the operational task on the system. Lack of communication generally results from incorrect identification of the information by the receiver (fixation). Focusing on performing the task within the technical system and/or complexity and situational efforts may lead to a breakdown of communication due to prioritizing the technical task instead of the communicative task on the receiver side. In this case, the receiver is no longer able to identify the meaning connected with information sent to him.

As an example, the receiver misunderstands information sent by the sender and mixes up the system to be operated (expectancy-driven processing). In cases of an overlap in tasks, in combination with a lack of experience, the receiver of information may understand the information sent wrongly (which leads to wrong decisions), especially if the sender gives an abstract input (assuming that the receiver may know what he means). Read-back or hear-back errors between controller and pilots in air traffic management belong to this category. They lead to safety critical events (Eurocontrol-Levelbust, 2004).

One of the most critical communication situations is information ignorance in order to achieve own goals. It is characterized by a reluctance of the information sender to transmit information to the appropriate receiver, even though the sender is aware that the receiver needs this information. The sender is no more able to imagine the needs of a potential receiver or lacks in this communication because this would mean to giving up an own goal. Such breakdown situations are quite common psychological effects. Also mobbing, the diminishing of the image of a colleague, belongs in this category.

Breakdown of communication is mainly due to a lack in co-ordination between receiver and sender (eagerness to act of sender combined with information ignorance of receiver). If several of the above-mentioned communication problems appear together, poor information exchange is the probable result.

In addition to these four cases, operational experience shows that good communication can be observed in high workload conditions (i.e., high objective workload). The assumption of a simple relationship, such as 'high workload leads to bad communication', does not explain this observation sufficiently. Detailed analysis of events shows that communication problems can be explained by the concept of cognitive coupling instead. If the operational task is in consonance with the communication task, good communication can be observed. If the operational task is in dissonance with the communication task, bad communication occurs. An example of such a well performed operational and communication task is the Davis-Besse Event.

Causes for Communication Failures

Going one step deeper into understanding communication failures, Figure 5.3 shows the relative importance of PSFs (Performance Shaping Factors) for verbal and procedural communication. The figure clearly indicates that verbal communication is always problematic if the communication partners are exposed to a certain degree of pressure (time-pressure or situational pressure, task-demands, task organization or preparation). On the other hand, procedural communication tends to fail more often due to ergonomic problems, like completeness of information in procedures, labelling or others. The factors mentioned in the figure refer to those discussed in Part I on the combination and interdependence of contextual factors (see Sträter, 1997/2000 for further discussion of the factors).

The figure shows that not only perceptual aspects (like precision of task or clarity of order), but also attitudes like willingness to send or openness to receive information in the current situation are important factors for verbal communication. Hence, verbal communication is clearly linked largely to the mental capacity and mechanisms to cope with situations than procedural-based communication.

Procedural and verbal communication can be distinguished in terms of types of errors and underlying reasons of failure. Several important distinctions can be inferred from the figure. Time- and situational pressure seems to affect verbal communication more than procedural communication. Verbal communication problems can be observed as having a stronger link to cognitive errors (i.e., leading considerably more to goal reduction). Cognitive aspects can be understood as essential for communication problems. This finding is of particular importance as failures of verbal communication have a higher potential for initiating events and hence should be considered in the Human Reliability Assessment.

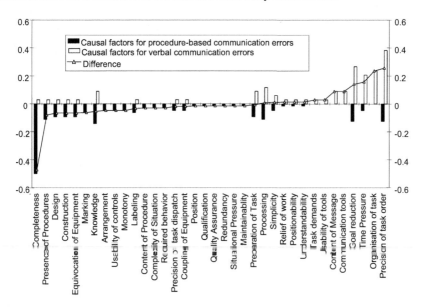

Figure 5.3   Profile of influences on procedural vs. verbal communication

Communication and Conflict

The cognitive control strategies, information ignorance, expectancy-driven processing, and fixation on processing can be observed as driving the communication behaviour (Sträter, 2002), as the following example shows.

One of the worst aviation accidents was caused by the misunderstanding of a controller instruction by a pilot on the Canary Islands. Two aircraft were involved in the accident, one from KLM and one from Pan Am. The KLM had been on a non-scheduled flight from Amsterdam to the Las Palmas airport in the Canary Islands, but had been diverted to Tenerife because of a bomb explosion in the passenger terminal in Las Palmas. Because of limited visibility and communications difficulties between air traffic control and the KLM aircraft, the KLM 747 started its takeoff while the Pan Am aircraft was on the same runway. All 234 passengers and 14 crew were killed in the KLM 747. Nine of the 16 crew and 321 of the 380 passengers on the Pan Am flight were killed (Airsafe, 2004). The communication failure was due to fixation on an established relation: the KLM pilots were under time pressure because they were about to exceed their maximum working hours allowed, if not departing as soon as possible, and mis-interpreted the instruction of the ground controller.

As the accident demonstrates, the four basic resolving mechanisms of the central comparator can be applied to the coupling of communication partners.

- Fixation on established coupling: Humans cannot apply their qualification and knowledge immediately and constantly in any situation. Especially in disturbances or under time pressure, where effective actions are required, knowledge –even of critical regulations– will not be applied optimally. As the example showed, communication may suffer from this as well. An information sender may be reluctance to communicate or may not provide information because he is busy with a specific task (e.g. problem solving).
- Information ignorance (utility-oriented): In some accidents, one will find that the workers communicated in a utility-oriented way. The utility can be of different nature. Shortening the permitted walking routes in order to save walking time is a typical example. Decisive for the utility one expects is the short-term and immediate utility within a situation – the long-term utility or the potential long-term damages are not considered in these cases. In the example above the information sent was not processed because the pilots were in utility-oriented information processing modes (they did not want to exceed the maximum allowed working time). The utility-oriented mode may result in communication avoidance on the side of the sender.
- Expectancy-driven (habitual): Relatively often one will find communication problems caused by the habits of the persons involved in the event. Working according to habits means having experience in how to deal with a system efficiently and productively. Therefore, we appreciate if workers have relevant experiences for a certain post. Experiences, on the other hand, let us fade out side effects, which we usually do not need to consider for accomplishing a

task. If these side effects are not considered in a situation where they should be considered, this may lead to a misunderstanding on the side of the receiver due to wrong assumptions about the information received. In aviation, the habitual effect of communication errors by misunderstandings is safeguarded by permanent variation of the shift constellation. Almost stable shifts are used in the nuclear industry instead. Both approaches have advantages and disadvantages. While the aviation approach is able to avoid misunderstanding due to habitual effects, it does not account for the efficient interplay needed in critical situations, which was the design criterion for the nuclear solution. A balanced job-rotation approach combines both advantages.

- Information and goal elaboration (rationale): Information or goal elaboration means all alternatives and combinations between the given information and goals are considered in the cognitive processing. In the case of information and goal mismatch, a communication partner may not communicate because he is busy with his internal world. However, human communication acts on a rather rational level. Behaviour and communication take more time and are consequently inefficient. That quantity of time is usually not available in work-situations. This is one reason why one can only expect in a minority of cases that a human acted rationally in communication.

The cognitive processing loop obviously affects both motoric behaviour and communicational behaviour. In respect to the communication, the internal goals play a vital role for group processes and conflict generation.

The four main properties are connected with fundamentally different strategies of how to deal with a person involved in the event and to find appropriate precautions. If, for instance, an accident happened because of mental utility, rational precautions (e.g. education on technical subjects) are largely useless. Accordingly, measures like training or instructions are probably not effective, because they do not fit to the human mechanism that led to the error. Depending on the mechanism that led to the erroneous behaviour, different approaches are necessary on how to communicate with the person involved in the incident.

Planning of communication strategies The situation of incident investigation can be understood as a conflict situation in the communication of the investigator and the investigated person or group. The classification of the cognitive processing loop helps the investigator to derive an efficient and right strategy for communication.

Consider events like the following. In aviation there might be a deliberate overlook of system signals in order to land the aircraft, as happened in the accident of a Boeing 747 in Kuala Lumpur (ICAO, 1989). Pilots ignored the (noisy) alarm of the Ground Proximity Warning System (GPWS) in order to get the aircraft landed. In occupational safety, this might be a deviation from permitted walking routes or the circumvention of two-hand operated machines to increase the efficiency in a piecework-based production cycle. In the nuclear industry, this might be the violation of independent treatment of redundancies in order to save working time. In Air Traffic Management, this might be a deviation from rules in order to increase efficiency. Such behaviour is often implicitly supported because it unhealthily combines the utility of the person who does the violation (e.g. to increase his or her salary or to gain acceptance in the job) with the utility of the management (increase of productivity). The nuclear accident of Tokai Mura was already discussed as a potential outcome of this unhealthy relation as was the mid-air collision of two aircrafts at Lake Constance.

A blame-free incident investigation needs to take the cognitive resolving mechanism into account that the person used during the accident or incident. Additionally, the resolving mechanism of other staff and management should be considered. From the analysis of the cognitive resolving mechanisms that were valid during an accident, conflicts and ineffective countermeasures can be inferred. For this purpose, Figure 5.4 illustrates the role of the resolving mechanisms in communication.

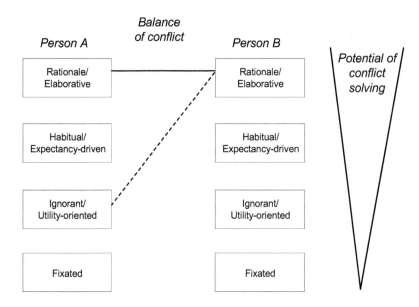

Figure 5.4   Balance of conflict

In all of the examples mentioned above, the humans made a mistake due to mental utility. If an investigator (person B in the figure) attempts to define countermeasures for the person who did the error (person A in the figure), the investigator usually starts on the elaborative level and assumes the person committed the error is on the same level. Consequently, he may come up with training procedures, for instance, in order to tackle the rational, elaborative level of cognitive processing of person A.

However, these countermeasures can only be effective if there is knowledge about procedures missing (i.e., if person A did the error in an elaborative cognitive processing). In accidents such as the ones described above, the humans are however acting in a mental utility. They were probably even aware of the procedures. Because they acted under mental utility, training as countermeasures is hence less effective. Note that the resolving mechanism can change from situation to situation and is not a latent trait of a specific person or group of persons.

Legal systems would deny the role of utility and consider humans in such situations as rational. Consequently, the persons committing the error have to be afraid of being condemned in case they cause an incident or accident. In addition, incident investigators treat humans in such situations usually as rational. This conclusion can be drawn from the means introduced to prevent an error from reoccurrence. Preventing a utility-related error by training, education or punishment has often little effect on the person, because they do not change the utility. However, such means increase the conflict between the investigator and the person who committed the error. The communication imbalance increases but the conflict is not solved, the same error will reoccur.

An investigator has to take into account the level of the resolving mechanisms, which were effective during the event. Suggesting training as a precaution to avoid reoccurrence will fail, because person B did not hit the valid resolving mechanism of person A. Person A will perform the training but will not change his behaviour because the mental utility will not change with training. On the other hand, if an investigator accuses a person as having acted in an utility-oriented manner during an accident, while the person was actually on the elaborated resolving-level, will lead to a wrong accusation and conflicts. In the worst case, the investigated person will stop the communication and refuse further investigations. Critical for conflicts are the utilities of communication partners. In case they fit, they provide group thinking, or form alliances, if the utility matches temporally. Different goals cause conflicts if the goals do not fit between the communication partners. This will be exaggerated if one or both of them become fixated on their goals.

Any imbalance in the levels is equivalent to a conflict with a disadvantage either for the investigator or for the investigated person. Countermeasures can become counter-productive or the information flow may break down. The figure can be used to avoid ineffective countermeasures or conflicts due to wrong accusations. Note also that an investigator may not be free of biases and may investigate in a utility-oriented manner although he should investigate in an elaborative way (he may for instance have a bad image about person A).

The ideal situation is that both the person who did the error and the investigator are on the elaborative level. Conflicts can be solved easily by building up memory (experiences or concepts) or by providing additional information (e.g. training).

With increasing distance from the rational elaborative level, the potential for conflicts increases. The same holds for the balance. The higher the imbalance between the different communication partners, the higher is the opportunity for a potential conflict between both and the lower the potential to solve a conflict.

Error investigation and prevention means taking into account the cognitive resolving mechanism of the person who committed the error. Table 5.2, based on Shell (2002) and Siemens (2002), maps the different communication strategies to the resolving mechanisms and mentions the most efficient means for each of the different resolving mechanisms.

Table 5.2   **Cognitive resolving mechanisms showed in events, most efficient communication strategy and most efficient means to overcome the situation**

Approach for ... ... Resolving mechanism showed during the incident	Analysis	Communication	Prevention
Elaborative (rationale)	Concentrate incident investigation on the external causes for the event	Start event investigation with causes outside of the person and classify individual error as due to external threats	Improve external working conditions, provide training
Habitual (expectancy-driven)	Investigate habits as a result of external threats and constraints	Start investigation with external constraints and trace habits back to the origins of how they were developed from the constraints	Provide counter experiences; Change external constraints while introducing costs for habits
Ignorant (utility-oriented)	Investigate potential utilities and costs as well as personal values and attitudes	Clearly state the utilities assumed (unmasking); elaborate alternative utilities and costs	Establish costs for the unwanted behaviour while establishing utility for the wanted behaviour. The utility needs to be free of contradiction with other potential utilities.
Fixated	Investigate habits and utilities as well as personal values and attitudes together with external constraints	Classify behaviour as typical human behaviour. Search for external constraints while also investigating the utilities. Give time for digestion and understanding	Improve external working conditions while establishing utility for the wanted behaviour

Because the resolving mechanisms are related to the emotional system, any conflict generates inherently strong emotional impacts of the persons involved. Only if the prevention fits the resolving mechanism, effective prevention can be achieved. This means that incident investigators have to align the strategy of querying facts about an event according to the resolving mechanism. A person, who made an error while rationally elaborating a situation, is the easiest candidate. The investigator can discuss the information considered and the intentions and means for prevention can be concluded. A person committing an error because of utility cannot be convinced with the same strategy. The goals and intentions have to be investigated and the means for prevention need to range from supporting the safety-related behaviour up to generating costs for showing unsafe behaviour.

Integrating Cognition into System-Organization and -Management

Trust as a Consequence of the Cognitive Processing Loop

A natural human behaviour following from the cognitive processing loop is to judge how the external world follows own actions. This holds for human–human as well as human–machine coupling. Once the partner fails in the functions assigned to him, the partner loses trust. If he manages even quite critical situations, he may well be over-trusted. Both situations are critical.

Distrust may lead to the rejection of an automated tool or a communication partner. The system or the partner will not be accepted. The expected improvement in the overall performance by introducing an automated system may not be achieved or communication will break down. This aspect is of considerable importance in the introductory phase of a new system where first experiences are made and trust develops. If the trust of the users in the system is not achieved it will take years until they will accept the introduction of the same or another automated system in their working environment (due to the sluggishness of the generic memory structure). The better an automated system works the more another problem comes into play. Over-trust in a technical system may be critical in situations where it would be better to distrust the system in order to prevent a critical failure (e.g. wrong resolution of a conflict by a conflict resolution system in air traffic management). The same holds for inter-personal conflicts.

As it is natural for a human to fail, any automated system may also fail in the designed functions. Incidents prove that over-trust is more critical for the overall safety of a system: in the Alitalia accident in Zurich in 1990, over-trust of the captain in his abilities and in the reliability of the plane was a contributing factor leading to the accident. The Chernobyl crew was the one that was most convinced about itself and its plant, before the accident. Considerable problems result from over-trust. Chernobyl showed that an entire industry may suffer from over-trust. After the accident, nuclear power was not considered as safe.

Bonini (2004) developed a conceptual model to understand the trust and links trust to cognitive ergonomics. She uses four levels: constructs, representations of a user in mental models, the primed decision-making, and the schemata level.

The approach reflects the cognitive processing loop by distinguishing concepts, experiences and coupling of internal and external worlds as essential elements. Trust is understood as consisting of a conceptual level (self-confidence, attitudes towards others), experience level (belief, expectations, constructs, patterns) and a control level (reliance, dependence, cooperation) that describes the coupling between two agents. Overall, this concept links important aspects to build up trust and provides a link of the trust issue to other existing ergonomic design methods via cognitive modelling approaches.

The approach understands trust as a relation between the internal and the external worlds in general and holds for both, the trust between the communication partners and between a user and a technical system. The more both worlds fit, the better the trust.

Transition and Change-Management

Trust is an essential aspect of introducing systems successfully (Kelly et al., 2003). Cognitive aspects in introducing new systems, procedures or organizational structures are, most of the times, sub optimal. Often, communication breaks down in critical re-organizations because cognitive aspects and mechanisms are not reflected well enough in the process of change.

As an example, air traffic management suffered from the issue of trust. The new control room in the upper area control centre of Eurocontrol could only go into operation after considerable delay due to the slow acceptance process, although the new system is now favoured by the controllers (Baret & Vermeiren, 2004).

The reason for such events is that the sluggishness of the human cognitive processing loop is often not recognized and changes are made too fast in order to achieve cost-efficient implementation. Too fast change processes, however, are only successful if they do not lead to considerable dissonance of the users. If the change process exceeds the users' concepts, experiences and coupling possibilities, the change may lead to resolving mechanisms of the users to cope with the changes. These may range from being curious about learning of the change, up to being strongly reluctant to change. Due to the sluggishness and hysteresis of the cognitive system, trust needs to be built up progressively and balanced between achieving curiosity and openness while avoiding ignorance and criticism, as Figure 5.5 outlines. Note again that the cognitive processing loop behind trust is strongly related to the emotional system. Hence, it is no wonder that ignorance and curiosity are highly emotional processes.

A balanced process to build up trust is an essential aspect for the successful implementation of new systems (Kelly, 2003; Bonini, 2004), because the cognitive level cannot be equalized with the external task a user should be compliant with. The successful introduction of systems needs to know the current level of concepts and experiences the users have. Data about these aspects can be obtained by knowledge acquisition tools (overview in Eberleh et al., 1989). These tools are developed to acquire detailed information about the underlying dimensions or concepts of human information processing (cf. Sträter, 1991).

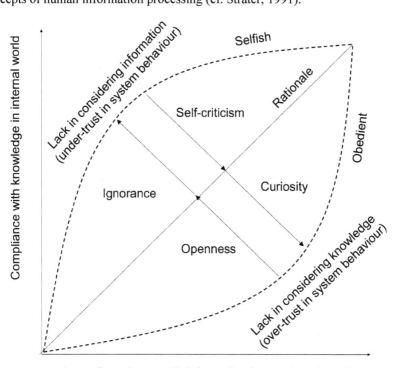

Compliance with information from external world

Figure 5.5   Progressive and balanced development of trust as a result of the sluggishness and hysteresis of the cognitive system

Risk Communication and Questioning Attitude

The mechanism of conflict by imbalance turns up as miscommunication in complex working environments (GIHRE, 2004). IAEA (2001a) investigated incidents in nuclear energy as documented in the Incident Reporting System (IRS) of the IAEA (International Atomic Energy Agency).

The report revealed that events are often insufficiently exploited regarding the various safety-issues of an event, because investigators as well as investigated persons are reluctant to communicate about safety-related events. They are afraid of legal proceedings, which turn their cognitive processing into a utility-oriented processing mode rather than a rational mode in order to protect themselves. The resolving mechanisms do hinder a rational risk communication. This lack of risk communication goes up to the level of those persons responsible for mitigating the causes of the events. Not all elements of the analysis are disseminated in the different organizations, departments, and divisions so that there develops a loss of knowledge and co-operation.

Such cases suggest that biases or pre-assumption lead to incomplete analyses of events. Such biases can be faced by providing resources for event information extraction (time, space) and by considering the information from different points of view by means such as extending the event analysis teams or evolving other departments into the evaluation process of the event.

The same report also reveals that a couple of safety critical events occur because the staff do not question processes with potential safety-relevance, because they are afraid to disturb the work objectives by either delaying the work or by appearing as being not well informed. Often a 'questioning attitude' is confused with a 'troublemaking attitude' in highly prescriptive or hierarchical organizations. Again, the communication turns into a utility-oriented mode. A better questioning attitude during the development of an event would have prevented the event. The events related to the questioning attitude aspect of this subject area deal with instances where employees were not properly briefed on a task and failed to solicit the required information. They also deal with instances where employees failed to apply knowledge, because either they were under pressure to complete a task or they thought the knowledge did not apply to the situation.

Lack of a questioning attitude often caused or significantly aggravated an event, such that the event would likely not have occurred or would not have been as serious if the people involved had asked relevant questions at the appropriate time. It is also related to deliberate violations of safety principles and procedures. These violations appeared to be tolerated by supervisory staff. Lack of a questioning attitude appeared to be related to lack of self-discipline and lack of enforcement. To improve performance in this area, the reasons for these violations need to be understood. These reasons include shortcomings in work administration or procedures. Enforcement of discipline alone will likely not succeed in eliminating these violations.

Organizations need to adopt an open and blame-free culture where staff can gain a feeling of ownership because both are necessary for the development of a questioning attitude by staff. Safety management systems request such open, blame-free culture (Balfanz et al., 2004). However, only if the cognitive mechanisms behind the human error are recognized and taken into account, can safety management systems be practised (Sträter, 2004b). The IAEA report hence concluded to:

- Encourage staff to develop a healthy questioning attitude by establishing mechanisms to support staff and minimize the amount of effort required to get answers. These mechanisms could include technical support, knowledge databases, access to event information, and clear communication channels. Staff should also be provided with opportunities such as team briefings or meetings to obtain clarification from both supervisory or management staff. Staff displaying a healthy questioning attitude should be encouraged by management through acknowledgment and openness.
- Minimize the number of violations and improve discipline by clarification of the required safety principles and procedures and of the potential

consequences of not adhering to these principles and procedures. This includes consideration and enforcement of principles and measures to ensure staff are motivated to follow the safety principles and procedures (e.g. clarification for what safety principles and procedures are necessary) as well as clear rules and measures for the non-adherence to discipline.

How important communication and trust are for the successful realization and 'living' of safety shows in the investigation of NLR, the Aviation Research Centre of the Netherlands, on the interrelation between different aviation players, like airlines, air-traffic-control service providers and regulators states (Roelen et al., 2003):

> The work of the FAA [US Federal Aviation Authority] Certification Process Study (CPS) Team is very relevant as this team examined interfaces between certification, maintenance, and operations. Among the conclusions of the CPS team are the following:
> Critical information may not be available to those that could act upon it. Organizational barriers to communication, failure to recognize the need to communicate, information overload, and language differences, may all contribute to information flow breakdowns.
> Significant safety issues learned through accidents are sometimes lost with time and must be re-learned at a very high price.
> Traditional relationships among the regulators and industry have inherent constraints that have, in some cases, limited the ability to effectively identify and act on accident precursors. Further safety improvements will require significant intra- and inter-organizational cultural changes to facilitate a more open exchange of information. Regulatory solutions alone cannot achieve the desired results.

The report of NLR shows a typical outcome of the cognitive processing in organizations. The system develops into an unsafe state due to loss of trust and preference of individual utility or economic preferences.

Safety Management

Safety management systems are currently established in any industry to build trust in safety and to overcome the effect of individual utility and economic preferences instead of safety targets in hazardous industries. However, all industries have problems in realizing safety management effectively. The safety management systems do not sufficiently consider the cognitive mechanisms of the staff that have to implement safety (Sträter et al., 2004b). Staff are quickly overloaded by the procedures envisaged for establishing safety. Safety becomes an additional task, which leads to constraints of the persons, compromises between safety and efficiency and eventually to decisions of the staff to do safety according to regulation or even to disregard safety. Current implementation of safety management systems provides a seemingly safe system, which can lead to an even more critical situation than no safety management system.

Safety management is therefore in a similar situation to software ergonomics. ISO-9141 suggests a strong participation of users in the implementation of software in order to assure trust in the approach and communication.

Eurocontrol, the European Organization for the safety of Air Navigation, therefore suggests in the future Air Traffic Management (ATM) safety strategy to provide resources to establish safety as a living practice (Eurocontrol-Agenda, 2005). Safety management tools and methods need to be practised in order to assure that safety is lived and not only documented. Safety practice means understanding safety in the constraints of the real operation of future ATM. To establish living practice, the operational constraints preventing safety from being practised need to be understood. In particular, the right timing for introducing safety-means or safety-measures at different ATM entities needs to be assured in order to minimize organizational deviances from safety targets.

Adequate management of all resources and the relationship within the ATM system between people, their environment and equipment concerning responsibility for aircraft separation, will be the main subjects for strategic safety improvement.

Since human performance is a crucial component of ATM, human involvement and commitment issues should be considered throughout the concept transition process. This is necessary to ensure commitment and ownership of all actors of the significant changes that will occur. The experiences from the nuclear industry have shown that loss of corporate knowledge may lead to serious incidents. Being aware of the competence level required for new ATM systems and maintaining existing knowledge is therefore an essential part of incident avoidance.

Organizations are often forced to make decisions under different considerations. Several objectives need to be met such – as in air traffic management for instance – safety, efficiency, capacity, environmental aspects and security. The role of decision-making is to be reflected in an integrated management system. The major task of an integrated management system is to resolve the conflicts between the different objectives like efficiency and safety but also safety and security. After 9/11, for instance, the cockpit door is required to be closed for security reasons but this has an impact on safety because pilots now need to communicate with the crew using telecommunications. Telecommunication, however, is the means to communication with the controller and hence an interference of the task to communicate with the controller is created, which has a safety impact. As a second example, an environmental objective of reducing noise in highly populated areas surrounding an airport may lead to a decision for more risky approach-routes on the political level.

An integrated management system therefore harmonizes and integrates several management systems in order to avoid jeopardizing safety practice due to potentially conflicting objectives of cost, quality of service, flight efficiency, or the need to alleviate environmental impacts (Eurocontrol-Agenda, 2005). Understanding of the cognitive resolving mechanisms of the persons involved is essential for practising safety.

Integration of Cognition into Design and Operation

The Remote Access to Cognitive Performance in Design

Cognitive task design requires a close link to the persons (or groups) being investigated in order to get the information about the cognitive processes during task performance. Usually, methods like rapid prototyping are used and potential user-groups are confronted with the design of the new system. Their performance is then measured according to constructs like workload or situational awareness. Cognitive task analyses are usually performed to support and to facilitate the process of prototyping. Various techniques have been developed to evaluate such investigations, like video confronting, eye tracking or various workload measures for instance (such as the workload measure NASA-TLX – the NASA Task Load Index).

For several instances of cognitive task design, such techniques cannot be applied, because the direct access or contact to the 'piece of interest' (the human being, or the group of persons, whose cognitive performance is being investigated) is missing. In such cases, none of the usual cognitive design methods can be applied with sufficient validity.

This is the case in various settings with a high interest in explaining and predicting cognitive aspects in order to derive design suggestions for technical systems. Among these settings are:

- Assessment of cognitive performance of humans in events or incidents based on evaluation of past experience where the access to the person(s) directly involved in the events or incidents is lacking.
- Assessment of cognitive performance in early design stages where an experimental setting has to be prepared, where a user group is not specified, or a detailed specification of the design of the technical system is not available so that none of the usual techniques can be applied.
- Assessment of cognitive performance in safety assessments during failures in the system, although the performance never occurred and probably will never occur.

Design and incident analysis have common objectives in the proper design of the system to avoid human error, but also common constraints in the accessibility of data on cognitive processes.

When an event or incident happens, usually no observer is present to investigate immediately the cognitive behaviour of the person(s) who performed the crew. In cases where the incident led to fatalities, as is the case in serious disasters, it is not possible to ask the person(s) afterwards about the cognitive processes, which led to such a bad consequence. Instances of such a situation are Chernobyl, Tokai Mura or the loss of the crew in an aircraft crash. Beside these rather public events, there are many 'daily' incidents or abnormal events recorded in several industries, which can be exploited for further system improvements.

Usually these are filed in written event descriptions, which may be enriched by further information, like voice- or video-recordings.

Information about cognitive processes may be lacking even if the users involved are available. Since investigations are delayed for hours or days, the persons interrogated are less able to reproduce the details of the cognitive aspects that occurred during the event. Even rapid investigation teams cannot investigate all incidents. Any cognitive analysis in these settings is therefore a remote analysis. Past experience, on the other hand, is a key information source for improving any type of system. New methods on Human Reliability Assessment, for instance, put great effort on getting information about cognitive aspects of human error based on incidents (OECD, 2000).

Remote settings also exist for designers not having the resources available to perform a prototype experimental study or for management not able to speak with their staff about all the concerns they have. The cognitive processes cannot directly be investigated (e.g. using video-confronting methods). However, the information about cognitive processes is extremely useful for improving the design processes or the conclusions from an incident investigation.

Often it is concluded from the fact that a person cannot be interviewed that no serious cognitive analysis can take place. This is one of the major criticisms for investigating cognitive performance from written incident reports and for introducing cognition into early design stages. Control modes as discussed at the beginning of Part III of this book are flexible to overcome this limitation. They are robust and generic enough to work in remote settings but allow a description and prediction of cognitive processes to a certain level of detail.

Whether control modes are a useful and valid approach for remote settings, either for describing cognitive processes or for predicting cognitive errors, was investigated in Sträter (1997/2000) using 220 incidents from nuclear power plants with human error contributions. The validity of the approach was tested by comparing the prediction of the causes for cognitive errors using the cognitive control modes with statements about cognitive efforts that were directly achievable from the incident descriptions. Figure 5.6 shows this comparison.

The rank-order of the information about cognitive errors provides the same predictive value in respect to the causes for cognitive errors. The importance was determined by calculating the frequency of an influencing factor related to the number of events in which a human error occurred. Positive values give the relative importance for high demanding coupling; negative values give the importance for low demanding coupling. Note that the influencing factors mentioned in the figure are the same as explained in Part I of the book related to Figure 2.7 regarding the dependencies of causes.

Based on the correspondence between coupling modes and direct information it can be concluded that the approach of systematic description of cognitive coupling between human and technical systems provides a basis for prediction of the cognitive control strategies and hence for predicting cognitive behaviour. The next section describes how the control modes can be used to improve the design of new systems by human automation management.

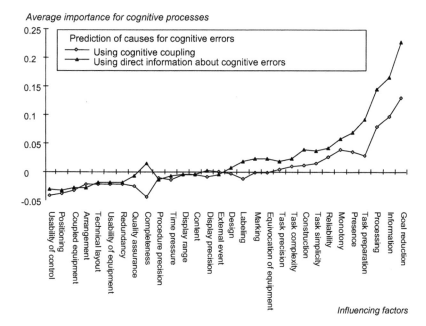

Figure 5.6   Profile of influencing factors in relation to cognitive demand

The Cognitive Control Loop in Dynamic Situations

Cognitive demand is not only related to the static aspects of a task. Cognitive demand varies with the dynamics of the situation. The discussion on the memory span (Part II) explains why. The memory span of 7±2 is only achieved during static tasks with sufficient time to elaborate and memorize the chunks. The amount of chunks to keep in memory is reduced under dynamic situations. The reduction is equivalent to the reduction of the number of decisions a human can perform per time (as represented in Figure 4.16, Part II).

Within the Group Interaction in High Risk Environment project, this was shown for a number of different industries and working environments (GIHRE, 2004). Pilots show a reduced set of communication patterns during critical situations. This could be validated in an extensive study on Crew Resource Management of pilots during simulator studies (Häusler, 2005). In the nuclear industry, it could be shown that experienced operators tend to compensate for high workload by reducing efforts to take new information into consideration (Fukuda & Voggenberger, 2004). The effect was less extensive for less experienced persons. Eye tracking analysis was used to show that information distracting from the current task or requiring a task shift is most likely neglected. Whether the information-processing mode is auditive or visual does not affect this tendency of excluding certain information from the processing loop.

A reduction of the memory span cannot only be explained by task interference and load of the working memory, as often stated in the literature (e.g. Wickens & Hollands, 2000). In a study on the use of navigation-systems during car driving, Rassl (2004) showed that the memory span is reduced. This has an impact on the driver's capability to process the depth of menus or other choice-related interactions. Schweigert (2003) investigated the compensation strategies that drivers chose to deal with different demands in different traffic situations, like driving on a highway vs. country road.

The driving task had to be performed under a given secondary task consisting of entering information into a device depending on choice reaction (either auditive or visual). He validated the existence of the cognitive resolving mechanisms in car driving and could identify a ranking in the resolving strategies as well. First, the search for new information is reduced. Seemingly irrelevant information is neglected and signs irrelevant for traffic are focused on significantly less or for shorter time (neglecting irrelevant information). Second, experience-based processing on relevant information is introduced and the duration of fixations on relevant information is reduced. This experience-based processing first optimizes time and, second, spatial orientation (optimizing search duration before neglecting relevant information). Third, utility of own goals is introduced combined with an increased reliance on rule compliance of others (reliance on rule compliance of others). Finally, the driver fixates the secondary task and omits traffic relevant fixations (fixation on additional task). Figure 5.7 shows the resolving strategies plus the observed proportions for different modalities (auditive vs. visual). The figures express the percentages of items neglected per each resolving mechanism, separated for auditive and visual secondary tasks distracting from the car-driving task.

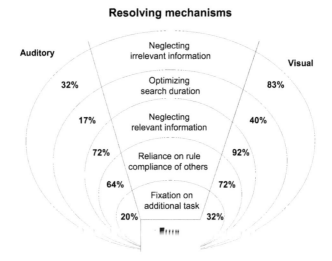

Resolving mechanisms

Figure 5.7   Resolving mechanisms used during car driving

Farmer & Jones (2001) validated the impact of secondary tasks on human performance for aviation in studies related to the party-line effect (the parallel listening of pilots to controllers' communication with other airplanes while flying their own airplane). Another study in air traffic management investigated the so-called head-up head-down problem. Airport operations require monitoring and assimilating of information across numerous displays. New technologies distract the controllers from fixations on the runway.

Human Automation Management

Virtually all industries are currently undergoing a huge increase in the use of automation. Increased automation goes hand in hand with an increase in the severity of accidents if the automated system fails but, on the other hand, may reduce the frequency of such severe accidents if the automated system is built reliably.

The success of such automated tool support will depend on the degree to which cognitive aspects of human performance are taken into account in the design and implementation of automation. In particular, automation may influence the cognitive performance and may result in changes to required skills (making some obsolete, whilst requiring new ones). Additionally whilst such tools may overcome some 'traditional' error forms, new errors may arise. Teamwork may be affected by automation, and the user must be able to recover in the case of an automation failure. There are also issues of trust and confidence in the tools that must be addressed to ensure full usage of the tools.

Automation leads to latent errors on the operational level if not designed according to the cognitive characteristics of the users. An observable automation issue, which led to an incident, is likely not mitigated. As discussed in Part I of this book, it is usually too expensive to mitigate design errors because they may require the entire design process to be re-iterated. Hence, even small modifications may be expensive and are avoided. This means that latent error forcing conditions and the risk that the user will make errors remain in the system and compensation means are not effective.

To overcome unsafe system design or unsafe system change management, potential latent failures in design have to be better analysed and predicted in order to anticipate future design phases at a time where the design can still be modified in a flexible way. It is necessary to have the means of determining which cognitive processes and activities are affected by certain design solutions and to get a feeling for the potential human errors implied by them. The project SHAPE (Solutions for the Human Automation Partnership in European ATM; Eurocontrol-Shape, 2005b; Sträter et al., 2004a) developed a framework for proactive cognitive task design. It aims to assist the designers of a system in order to prevent human errors before they occur in incidents. Figure 5.8 outlines how this general framework would predict cognitive processes for the Air Traffic Management Environment.

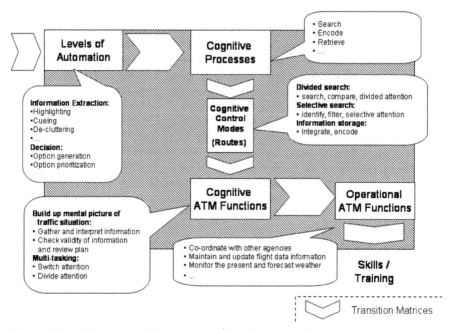

Figure 5.8   The general framework for skill set prediction in SHAPE

In order to determine how an automated system can affect the cognitive demands of a controller, the framework starts by classifying a system according to its levels of automation (Parasuraman et al., 1993). The levels of automation describe the functional support that the automation aims to provide to the controller. By establishing the links between the functional support that the automation aims to provide and the cognitive processes of the controller, it is possible to predict the impact of the automated system on cognitive processing stages. By mapping the cognitive processing stages to the cognitive ATM functions and the operational ATM functions, it is possible to identify the controllers' mental demands and the skills that will be impacted by the automated system and so need to be trained.

A critical link in the framework is the link between cognitive processes and cognitive functions. This link is established using the cognitive control modes (called *Cognitive Routes* within the SHAPE project). They are the essential bindings to tie the rather cognitive concepts of levels of automation and information processing with the ATM specific classifications of cognitive and operational ATM functions.

The approach of the control modes allowed the combination of a variety of different cognitive approaches using transition matrices. Herewith the framework integrates existing approaches on assessment of workload or situational awareness with ATM-related classifications of cognitive functions and training methodologies. The combination allows identifying the impacts of automated systems on controllers' cognitive processes and enables an integrated prediction of:

- The impact on controllers' team-working, trust, and situation awareness as well as problems related to HMI design (via the link of levels of automation and cognitive processes).
- The impact on controllers' potential to recover automation failures (via the link of cognitive processes and control modes).
- The impact on controllers' mental workload via the link of control modes to cognitive ATM functions).
- The change in controllers' skill requirements based on how the controller will function when using the automated system (via the link of cognitive and operational ATM functions).

Practical use of the framework for skill prediction Comparing experimental results with the predicted skills tested the validity of the framework. Automation on the data-link between pilots and controller was utilized for the validation exercise.

Without exception, all predictions made using the skill prediction process of the SHAPE Framework were actually observed by at least one of the seven controllers participating in the trials. Nearly 75% (26/35) of the predictions made were observed by four or more of the controllers (i.e. more than half). These findings are illustrated in Table 5.3, showing the percentage of controllers (out of seven in total) who observed the change predicted (part A of the table). Negative percentage states how often a new skill was noted as relevant but not predicted (part B of the table).

The results suggest that the prediction process does work and is fairly accurate with few if any predictions that were false positives. However, the controllers made additional observations that were not predicted (part B of the table). This would suggest that the process still has uncertainties, as any prediction has. However, the results lead to internally consistent results (i.e. the framework does not under-predict). Further details on the study and predictions can be obtained from Sträter et al. (2004a).

Use of the framework Using the framework enables the designer to see which cognitive processes and functions are affected by the system's design. This enables earlier assessment of design options and/or comparison of different systems as well as earlier alignment of training needs and system design. In addition, the various steps of the framework allow users with different backgrounds to attach their methods to the predictive framework. For instance, the framework can be linked to workload measures or to error-prediction tools (Shorrock & Sträter, 2004). It can also be linked to the Eurocontrol incident analysis tool HERA (Isaac et al., 2004). Hence, the framework can be fed with existing incident data so that an assessment of potential errors can be made. Herewith, a designer may get an insight about what types of errors may be introduced in given contextual conditions.

Table 5.3 Overview of the skill changes predicted and observed in the data-link experiment

A) Skill elements predicted (ATM functions; seven controllers)	%
Use standard phraseology and apply standard speech techniques to provide a service and to issue relevant clearances/instructions.	100
Obtain/verify read backs and detect/correct read back errors.	100
Switch attention, divide attention and/or time-share between tasks.	100
Incorporate new flight strips into own flight data display.	100
Reflect new information and changes affecting traffic situation and/or diagnosis of problems in the strip-marking and strip arrangement correctly.	100
Keep own strip display sufficiently tidy and meaningful.	100
Understand the coordination requirements, coordinate with other agencies/ATS centres, and respond to telephone calls with identity.	86
Maintain accurate flight data display (paper and/or electronic) and update traffic information and data on all aircraft after issuing any clearances / instructions.	86
Record all coordination outcomes.	86
Move the flight strips and obtain an effective strip arrangement for planning.	86
Support and share tasks effectively and/or offer assistance to team members.	86
React to conflict information and/or conflict alert warning as soon as possible.	71
Issue appropriate and safe clearances and instructions (including diversions messages) to the pilot and/or relevant ATS centres/agencies using correct communication techniques.	71
Initiate and carry out coordination correctly and efficiently.	71
Accept and initiate radar handovers and identification efficiently.	71
Carry out efficient task management.	71
Prioritize requests and only accepted requests, which ensured efficient workload (his/her own and/or team's workload) was maintained.	71
Prioritize own tasks effectively and efficiently and react promptly upon receiving relevant information or requests from pilots and other agencies and use the correct approved operating procedures.	71
Use the R/T and telephone equipment effectively.	71
Request assistance when appropriate and communicate needs clearly to other team members.	71
Identify, prevent, and resolve conflicts.	57
Evaluate options, priority of actions and consequences of the plan against the existing and anticipated traffic situation/conditions.	57
Ensure separation standards were achieved or maintained, using separation procedures, given changes in traffic situation.	57
Ask for relevant information from the appropriate persons and pass on relevant information to the pilot/other team members/other agencies as necessary.	57
Monitor and actively check the relevant display/external information to gather the appropriate information for the task.	57
Take action to correct mistakes or ambiguities of other team members' actions.	57
Execute revised plan given significant changes (including impact of weather situation) to traffic.	43
Allocate correct priority according to the category and type of the flight (including Airspace Reservations/Non standard flights).	43
Navigate around workstation displays (including communication panel) effectively and efficiently.	43
Take correct action (including identifying traffic and verifying mode 0) to ensure that flight plan data are complete and valid.	43
Enable aircraft to comply with the flight plan and assist efficiently in navigation of the aircraft as required (including diverting, non standard, special flights)	29

**Table 5.3   Overview of the skill changes predicted and observed in the
data-link experiment (continued)**

A) Skill elements predicted (ATM functions; seven controllers)	%
Accept feasible and relevant requests.	29
Provide an effective service (includes radar advisory, control and information).	29
Offer assistance to the aircraft when appropriate.	29
Provide a service within the area of responsibilities (including elements of airspace delegated by other ATS centres) that is safe and efficient to the category of airspace.	14
B) Skill elements not predicted but observed (ATM functions)	%
Diagnosing problems and implementing solutions.	-43
Make decisions/solutions for the traffic situation/conditions.	-57
Perform the necessary planned actions before aircraft arrives into the sector or into area of responsibility.	-57
Validate and verify Mode A & Mode C using correct R/T.	-57
Disseminate weather information.	-71
Raise warning or blocker strips and move strip into display.	-86
Reflect the plan and status of tasks in the strip-marking and strip arrangement.	-86
Use notes and physical cues to put in place effective mental reminders.	-100

The ability to identify the impact of automation on Human Factor issues such as training needs, mental workload, situation awareness, etc can aid design decisions, simulation planning, training needs assessment and successful transition training. Being able to determine skill changes as a result of automation early (i.e. in time to influence design options and cost-benefit trade-offs) can provide the opportunity for design, training and selection to become more proactive to technology change. The skill prediction process is principally intended for deployment early in the system development but it also enables skill predictions later, in either real-time simulations or implementation trials of an ATM system.

A unified, systematic approach to analyse automation and understand its impact on controller mental processing, functions and skill requirements enables not only comparisons between different design options or automation systems, but also ensures commonality between designers and analysts. It can also provide a traceable process that can be re-used either in iterations throughout the life-cycle or for different systems. This includes:

- A common understanding of engineers, software designers, controllers training specialists, etc about issues related to future technological developments.
- A mapping of generic ATM functions (and by inference the mapping of skill requirements) to technological developments.
- An efficient and responsive way of addressing skill changes to future technical assumptions and developments.
- An assessment of inter-dependencies between different Human Factor-related issues in automation, like workload, situational awareness, team interaction, trust, or human error.

The Link of Retrospective Analysis and Prospective Assessment of Human Error

The architecture of the cognitive processing loop implies the importance of knowing about the experiences of the persons in order to understand their behaviour. Experiences and concepts used in reasoning built the generic memory architecture. Experiences explain how persons practice under certain daily conditions that may result into critical situations and require resolving mechanisms. Resolving mechanisms follow if dissonance occurs in order to achieve consonance with the external world.

Experiences are, according to the cognitive processing loop, essential to understand human behaviour and error. The importance of understanding human behaviour based on the experiences and constraints of persons is also highlighted in the Functional Resonance Accident Models (FRAM; Hollnagel, 2004). A key question is: how can one collect humans' experiences in order to use them for assessment and hence efficient accident prevention?

One can think of a number of sources to evaluate human experiences, ranging from direct interrogation of workers using knowledge acquisition techniques, experiments or operational experiences (e.g. Eberleh et al., 1989).

Understanding human cognition needs to represent the experiences the humans gained in their working environment. Operational events have a unique advantage compared with other sources. They show daily routine, the way people work and understand their working environment, which experiences they gained and use in the real working environment, and which biases they have. Compared with the alternative to inquiring into peoples' subjective statements or putting them into a more or less virtual experimental environment, operational events are the most objective source for understanding the experiences. On the other hand, they are often argued to be too far away from cognitive analysis, but the approach of cognitive control modes was concluded to allow the evaluation of this remote access to cognitive performance as stated in the previous chapter.

Events show the importance of Human Factors for the safety of technical systems. In industries such as aviation and Air Traffic Management (ATM), the portion of Human Factor related events varies around 70% to 90%. ATM, with its considerable reliance on human interactions, belongs to the group of industries where unsafe events are related even stronger to Human Factors.

Often, Human Factor related events are directly related to the term Human Error. Although the term *Human Error* is a common one, it unfortunately implies that the cause for the event was due to an insufficient interaction of one or several humans with the technical system. However, the term *Human Error* should comprise all elements of the technical and organizational conditions that led to a certain, insufficient interaction of a human with the system.

The question for design and mitigation of events is whether the error is of a random or systematic nature (Rigby, 1970). Often, one can read in incident reports that the human error was singular and an accident happened because of unmanageable, coincident circumstances. However, after a collection of a couple of events, one suddenly perceives a systematic pattern behind the seemingly unrelated events. Often this systematic is not derivable from the classifications chosen for event analysis (the taxonomies) but from the pattern of causes or constellations.

In analysis of human behaviour, the analysis has to be related to human behaviour and not to technical components. In practice however, human errors are still treated separately for each discipline. A pilot's human error is not compared with the human error of a nuclear operator for instance, although many parallels on the basis of the cognitive processing loop can be found. As a consequence, error mitigation strategies are also more related to the enhancement of the technological barriers than to the enhancement of human performance in almost any technical discipline. An example, how little the mitigation measures reflect human error mechanisms is given in Sträter (1997/2000, p. 210).

An event analysis method therefore needs to have a detailed classification that has several taxonomies for errors, context, cognitive processes and behaviour. This requirement trades-off with the requirement of valid and consistent event analysis. Therefore, it was not systematically introduced. So far, little validity was assumed for such detailed analysis methods. By a systematic evaluation of variability between incident analysts and event types, Linsenmaier (2005) showed that validity can be achieved by computerized event analysis support. He developed an Event Organizer (EVEO) allowing the detailed analysis of human behaviour in events while assuring and monitoring validity in the event description. EVEO can be used independently in any industrial setting (nuclear, aviation, etc). Details can be obtained from EVEO (2005).

The Event Organizer EVEO was developed to support the retrospective analysis of single events by a computerized tool (Linsenmaier, 2005). The evaluation of a collection of events regarding their commonalties is performed by the CAHR method.

The CAHR Method

CAHR (Connectionism Assessment of Human Reliability) was developed to combine event analysis and assessment and therefore to base human reliability assessment on realistic experience (Sträter & Bubb, 1998). It follows the steps of the cognitive processing loop and investigates events regarding:

- Constraints (contextual conditions).
- Control modes (cognitive demand).
- Resolving mechanisms (consonance achievement).
- Behaviour (resulting error or good performance).

CAHR takes advantage of the connectionism modelling of cognitive performance. The term *Connectionism* was coined by modelling human cognition on the basis of artificial intelligence models. Connectionism is a term describing methods that represent complex interrelations of various parameters (known for pattern recognition, expert systems, modelling of cognition). By using the connectionism idea, the CAHR-method attempts to consider that human performance is affected by the interrelation of multiple conditions and factors (of internal as well as of external nature) rather than by independent singular factors. By this, it enables us to represent and evaluate dependencies of influencing factors and context of situational aspects on the qualitative side. It further suggests a quantitative prediction method that considers the Human Error Probability (HEP) as always driven by human abilities and the difficulty of the situation. Currently, the method is under further development for analysing events (see IAEA, 2001b) and for applying it to other fields of Human Factors (Linsenmaier, 2005; VDI, 1999; Bubb et al., 2004).

The basic idea of the approach for event evaluation and data collection is a detailed analysis of the information flows that were important in an event. Figure 5.9 provides a general overview about the event analysis procedure. Besides ergonomic and organizational aspects, the framework also allows to investigate the communication aspect.

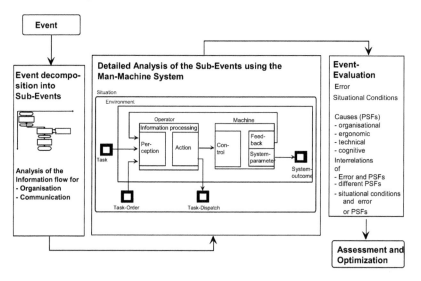

Figure 5.9   Overview of the CAHR method

Several applications in the past showed the value of the method, especially in the field of cognitive error investigation. CAHR was developed at the Gesellschaft für Anlagen- und Reaktorsicherheit (GRS) in the years 1992 to 1997. The database of GRS, with approximately 5000 events at German Nuclear Power Plants, was taken as the input source for the investigation. 232 have been selected for the detailed analysis. The 232 events could be further broken down to 439 sub-events (i.e. each event containing 1.10 sub-events on an average).

Complex working environments are modelled by several Man-Machine Systems (MMS), connected by the communication branches task order (receiving of information from another person) and task dispatch (providing information for another person). They represent the basic elements of the sender-receiver model within the MMS. Such decomposition also allows reflecting all operational levels (working, maintenance, design, organization and regulation) properly in incident analyses.

Figure 5.10 shows the general approach of how decomposition of complex working environments into MMSs may take place (Sträter, 1997/2000). The computerized incident tool EVEO supports this construction and analysis of the event flow (Linsenmaier, 2003). This decomposition into a time-person diagram is a common one and is also used in other incident analysis tools (for instance, Eurocontrol-Tokai, 2003; Wilpert et al., 1998).

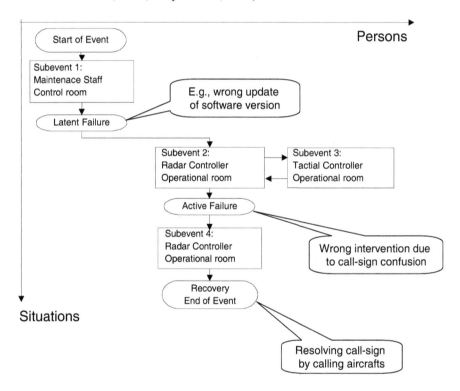

Figure 5.10   Event decomposition

The event decomposition shows the information flow and communication within an event. It therefore represents the dependencies between different persons involved in the event, as either persons initiating the event (latent or active) or persons involved in potential recovery actions.

In the hypothetical event outlined above, a maintenance person has wrongly updated software versions for the FDP system (FDP – flight data processing). The radar controller, in co-ordination with the planning controller, performs his normal activities, but not being able to identify the latent failure, the failure turns out as an active one. Finally, the radar controller recognizes the problem and resolves it successfully by calling the aircraft involved in the call-sign confusion.

Semantic Coding of the Experiences Represented in Event Information

To fill the generic memory structure based on events and to find the important information about errors and influencing factors, the following questions must be asked for each aspect within the MMS:

- What was the object of interest, e.g. 'valve'?
- What was the performed action, e.g. 'open'?
- What was the error, e.g. 'omitting' or 'too much'?
- What was the influencing factor, e.g. 'bad labelling'?

These questions provide information about the *Factual Aspects* of each MMS-aspect (the object, the action, the errors, and the influencing factors). According to epistemology, in Keller (1990) these steps were consequently called the phenomenological view (assignment of an error, step 3), causal view (relationship of step 1 and 2 to step 3) and actional view (step 4). A fifth factual aspect called 'element' was added in order to be able to describe the event in more detail and to make comments on the previous columns. This distinction is of importance because each step means a deeper insight into reasons for the error within the observation. The derived answers concerning the MMS-aspects and the description-aspects may be compiled in a table, as outlined in Table 5.4.

Answering all these questions (description of all factual aspects for every stage of every MMS that were identified within an event) leads to information about the context of an event. Context is defined here as

- the task-order and tasks,
- the information from the system to perform the tasks (feedback),
- the characteristics of the situation (e.g. time constraints) or of the technical system (e.g. dynamics), and
- the interrelationship of various MMSs (the operators, the management or organizational staff).

All these aspects together build the error situation or error context. In total, one finds plenty of specific questions that may be of relevance to describe the context. Overall, the table reflects the experiences of an operator in a hypothetical event.

Table 5.4   Illustration of a sub-set of experiences as coded in the CAHR method on the basis of a hypothetical example

Compo-nent	Sen-tence	Object	Action (verbal)	Indica-tion (error)	Property (PSF)	Element (Sub-generic term)
: (description of preceding components) :						
Task	1	Valve				Type X
"	"	"	Open	Omit	Time Pressure	High
Person	2	Operator				
Activity	3	Valve				
"	"	"	Opened	Too much		
"	"				Labelling	Legibility
"	"	"			Labelling	Insufficient
"	"	"			Legibility	Poor
:						
Feed-back	4	Message				Position-indication
"	"	Position-indication	recognize			
"	5	Display				Pressure gauge
"	"	Pressure gauge	Read off			
: (description of succeeding components) :						

A Connectionism Approach for Data-Representation

A connectionism algorithm was developed, which reflects the generic memory architecture as elaborated in Part II of this book. The approach represents the concepts and experiences as well as their interdependencies, and the relationships to contextual and situational conditions. CAHR represents therefore some kind of experience-based representation of cognitive performance. This information includes cognitive coupling, behaviour and errors, as well as situational conditions and causes for errors. The knowledge base of this model is subsequently built up by the systematic description of human performance and human errors. The level of resolution of the method is similar to the flexibility of describing events in a free text form. However, it is systematic enough to be elaborated systematically regarding statistical criteria and interdependencies between the terms used.

The connectionism approach of CAHR facilitates qualitative and quantitative analyses of the data collected. As a result, it becomes possible to deposit, in a uniform database, information for the evaluation of human reliability and for the optimization of the technical system. It also makes it possible to interrogate interrelationships of random concepts within the data structure (for example, relationships of errors and PSF; see Figure 5.11). As the connectionism approach represents the generic memory structure, the algorithm to maintain the data can be understood as some kind of language retention and production system. It contains language elements and represents different semantic elements on various syntactic layers. The procedure is therefore a language processing approach.

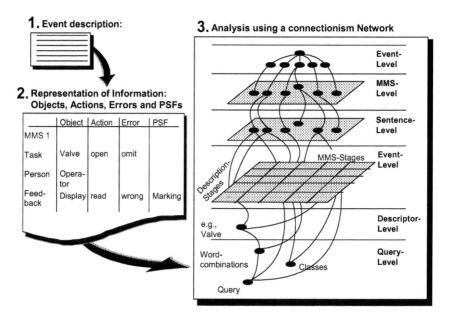

Figure 5.11   Overview of the semantic processing of the event information

Detailed Analysis of Human Interventions

The information flow does not provide detailed information about the working environment of the person (or group of persons) playing a role in that sub-event. For such detailed analysis, other approaches are required. A couple of methods were developed, like in the Air Traffic Management (ATM) environment: HERA (Isaac et al., 2004; Sträter et al., 2002) or TraceR (Shorrock, 1997). These methods are related to the operational level. From other industries a range of other methods are known (for chemical event analysis see Kanse and Van der Schaaf, 2000). The incident investigation tool HERA uses flowcharts for the detailed analysis. Flowcharts are one approach amongst many. CAHR uses tables. Independently of which syntax is used in the descriptive framework, the semantics and the general structure correspond greatly. Therefore, the technical content of the different methods is very similar and the methods can exchange well the information collected. This interchange-ability comes with the advantage of sharing data between the users of a particular method.

The detailed analysis is performed backwards from behaviour (including error types like omission or commission) via resolving mechanisms (like fixation) to control modes (like monitoring) and finally to constraints (contextual conditions like quality of working equipment etc). Figure 5.12 describes the steps to analyse the different contributors to an incident using CAHR.

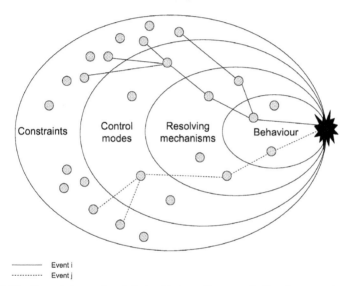

Event i
Event j

Figure 5.12   Event trace back into history from a defined error type to the underlying constraints

Operational Events as a Source to Represent the Experience Layer

Often, retrospective and prospective methods are discussed as being of a different nature. This is certainly true if one focuses on the time line they are dealing with. Retrospective analysis is dealing with events that have happened, and follows the time line backwards from the event to the underlying causes. By tracing back an event one is able to find specific causes or context constellations that resulted in the event.

Prospective assessment attempts a prediction of possible futures. Therefore, there is not one specific path (or pattern) that is investigated but several possible paths into the future, which may develop from a certain set of situations (Figure 5.13) to probable behaviour.

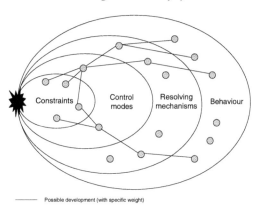

Possible development (with specific weight)

Figure 5.13   Possible development of a situation into behaviour

Retrospective analyses, tracing back of an event and prospective assessment, have differences. Retrospective analysis always shows a specific path into history while prospective assessment investigates possible paths into the future. Therefore, the direction of analysis and the resulting search or analysis scheme may differ between prospective and retrospective methods. In addition, the set of possible relations differs from retrospective to the prospective view. While, in a retrospective view all the relations between the behaviour and resolving mechanisms are, in principle, at hand, there are always several possible futures in the prospective view. In terms of data processing we therefore find a 1-to-many relation of behaviour to constraints in the retrospective view while we find a 1-to-many relation of constraints to possible human behaviour in the prospective view.

This difference has an impact on the analysis but not on the cognitive model used. Retrospective and prospective methods have common elements that are essential for analysis and assessments and that are vital for exploiting past information for prediction.

The most important common element is the common structure from behaviour to constraints and the common language (taxonomy) for describing either analysis results or assessments. The Figures 5.12 and 5.13 show the distinction of the CAHR (Connectionism Assessment of Human Reliability) approach using the levels 'behaviour', 'cognitive control', 'resolving mechanism' and 'constraint'. The Human Error in Air traffic management (HERA) approach for retrospective analysis of incidents and prospective assessment of human error distinguishes 'error type', 'error mechanism', 'information processing', and 'contextual condition' (Isaac et al., 2003). CREAM (Cognitive Reliability and Error Analysis Method) uses an analysis and assessment structure from 'action failure', 'cognitive function failure', 'cognitive demands profile', 'control mode' to 'common performance conditions' (Hollnagel, 1998). ATHEANA (A Technique for Human Error Analysis) distinguishes between 'error forcing context', 'error mechanism' and 'unsafe act' (NUREG-1624, 2000).

Generally speaking all the approaches have apparently a similar underlying structure. Both the retrospective view and the prospective view link errors to contextual conditions using a path from causes' via 'cognitive behaviour' to 'behavioural effects'. No matter which common basic elements one chooses, the language (taxonomy) for the event description and the assessment have to be the same, compatible or at least comparable if one wants to have a chance to make reliable and valid predictions. The elements of an error prediction in a fault-tree, for instance, should be coherent with the information collected in retrospective analysis. Any distribution is then the result of a collection of more or less complete historical data in a common language.

Hence, past events may be used for prediction if the appropriate cognitive model is underlying the transfer from retrospective behaviour to prospective prediction. Operational experience can be used to generate valid estimations on human mechanisms in operational conditions. According to Figure 5.14 this requires the same taxonomy (e.g. to transfer the error mechanisms and a valid cognitive model.

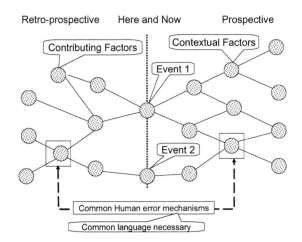

Figure 5.14   Necessity of a common language if retrospective analysis should be used for prediction

Integration of event data into human automation management design and organization of systems   CAHR does not restrict analysis to the sharp-end level of operation and allows inclusion of other levels involved in incidents, like design, maintenance, organization and regulator. This basic feature allows using operational experience not only for developing appropriate human reliability models but also to use operational experience to enhance human automation management further. Operational experience herewith can augment a human automation system as discussed in the previous chapter. In order to overcome Human Factor problems with the potential to lead to serious events, two essential elements have to be released:

- Feedback loops from operational experience into preceding design stages. These are design strategies and recommendations, rapid prototyping, and validating experiments in the product development phases and event reporting schemes in the product implementation phases.
- Predictive elements for the appropriate design and implementation of technology, like design decision support systems in the product development phases and Human Reliability Assessment (HRA) methods in the product implementation phases.

Integration of event data into the organization of systems   The systematic evaluation of operational experience allows finding the critical hurdles for system implementation related either to managerial tasks or to the work organization. Dealing with organizational issues means preparing the organization for the changes of systems or organization. Often, mistakes are committed at the organizational level because the principle of 'just implement' is followed. As a result, staff are often ill-informed about the planned changes, and systems are often perceived more as an additional constraint than a help in the daily work. Operational experience is able to provide information on the constraints staff are working under and to provide a strategy in implementing changes or new systems for the organizational level.

Operational experience can further enhance safety culture by generating a good co-operation and understanding among all operational levels (working, maintenance, design, organization and regulation). Such an enhancement would require abandoning the blame aspect of human error and take operational experience as a platform of good co-operation. Two preconditions needs to be mentioned for a successful use of operational experiences for this purpose:

- Operational feedback needs to be established on all operational levels so that no level is excluded from analysis (also in terms of legal responsibility).
- A cognitive model needs to be in place that avoids the blame aspect on human behaviour as done by the techno-morphologic information processing models.

Chapter 6

Assessment of Cognitive Performance in Safe Operations

Safety assessments are common approaches to judge the risk of complex systems. In order to include the human side, these assessments use Human Reliability Assessments (HRA) methods. Regulatory bodies are typically customers for safety assessments since they have an interest in the overall safety of the systems they have to regulate. In addition, assurances or industries use safety assessments if the safety status of a certain system needs to be elaborated.

Modelling and assessment of cognitive performance are major challenges in human reliability and system safety. This chapter describes how the cognitive control loop can solve actual problems of including cognitive aspects into safety assessments. It will elaborate on:

- A historical overview of the assessment of the cognitive reliability.
- The integration of cognitive performance and error into safety assessments.
- The assessment of the cognitive performance in safety assessments.
- The quantification of the elements of the cognitive processing loop.

A Historical Overview of the Quantitative Assessment of Cognitive Performance in Safety Assessments

Many different Human Reliability Assessment methods have been developed over the last decades. The methods can clearly be distinguished into 1st and 2nd generation HRA methods and a set of methods developed in the transition period between both. Further discussions about the spectrum of HRA methods can be found in (OECD, 2000; OECD, 2001; OECD, 2002; Sträter, 2004b).

Human Reliability Assessment in the 1st Generation

First generation HRA methods still remain in the techno-morphologic age of cognitive modelling and have little or no means of assessing dynamic situational factors. A wealth of literature can be found on 1st generation HRA methods that can be used for the quantification of human actions. Summaries can be found in, among others, Swain (1989), Reason (1990), Reer (1995) or VDI (1999). Further details regarding the information requirements of different HRA Methods can also be found in IAEA (1990). Essential 1st generation HRA methods are:

- THERP (Technique for Human Error Rate Prediction).
- HCR (Human Cognitive Reliability Model).
- SLIM (Success Likelihood Index Method).

This selection also represents which sources are typically used in HRA procedures for data concerning human reliability. Generally, there are four sources of data in HRA: estimates by experts, simulator studies, experiments, as well as operational experience. In the SLIM, the data on human reliability are based on the estimates of experts; in the HCR method, the data are based on simulator studies; in the technique for THERP (and ASEP), they are based mainly on operational experience and experiments.

THERP (Technique for Human Error Rate Prediction)   THERP is the most recognized method for assessing human actions in safety cases. It offers some flowcharts for the quantification procedure, and tables with predefined error probabilities for certain human interventions. The human is considered as a system component and is treated in a similar manner to technical components. This analysis is broken down into the following steps:

- Determination of system functions, which require personal actions.
- Task analysis concerning human actions, which can impair system functions.
- Estimation of probabilities of human error in these actions.
- Identification of weak points and derivation of recommendations for improvements to reduce the error probabilities in a new evaluation.

Within THERP, the so-called HRA action tree is used for estimating probabilities. Figure 6.1 shows a HRA action tree according to Swain & Guttmann (1983).

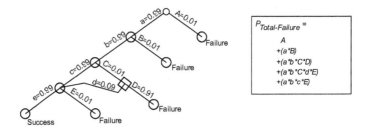

$$P_{Total-Failure} = A + (a \cdot B) + (a \cdot b \cdot C \cdot D) + (a \cdot b \cdot C \cdot d \cdot E) + (a \cdot b \cdot c \cdot E)$$

Actions
A/a: control-room operator omits to order the following tasks
B/b: Shift-personal omits the check of switch MU-13
C/c: Shift personnel omits the check or opening of pressurizer valves
D/d: Recovery by Shift-supervisor
E/e: Shift personal omits to lock pressurizer compartments

HEP-Probabilities from Tables
A=0.01 (0.005-0.05)
B=0.01 (0.005-0.05)
C=0.01 (0.005-0.05)
D=0.91 (0.5-0.99)
E=0.01 (0.005-0.05)

Figure 6.1   Action tree and error probabilities

The HRA action tree represents the sequence of necessary subtasks of personnel, which were determined by a task analysis. In order to determine error probabilities for the subtask, each subtask is classified according to the generic nature of the task (like perception or manual control) and essential PSFs (like ergonomic layout, time available or stress level) and then compared with predefined tables containing error probabilities for a comparable nature of the task or PSFs.

The error probability identified is then used in the HRA action tree to determine the overall probability of a task sequence. Finally, the HRA action tree is systematically examined for the possibility of so-called recovery factors and conditions of dependence between subtasks.

HCR-Human Cognitive Reliability Model   HCR, developed by Hannaman & Spurgin (1984) exploits the experience of experimental error research, which demonstrated that error decreases with the increase of available time (Wickens, 1984). In psychological literature, this relation is also known as the speed–accuracy trade-off or as 'Fitts' Law' (Fitts, 1954). This relationship is represented as a time-reliability curve (TRC) within HCR. The probabilities of an error in a given time budget are determined from simulator experiments with operating personnel. The determined probabilities were classified according to the cognitive levels of Rasmussen and Reason into time reliability for skill-based, rule-based and knowledge-based behaviour in order to reflect different cognitive demanding situations.

The HCR model received a systematic validation study that attempted to confirm the TRC. However, a validation study performed by Moieni et al. (1994) could not confirm the approach. It could also be shown that time is related to short-term reliability but it is not a sufficient parameter if the time available is high (Sträter, 1997/2000). Spurgin (2004) confirms not using the method further and to consider it more as an intermediate step of scientific progressing.

Despite this falsification, HCR is still used in some safety cases. This violates the rule of scientific reasoning stated in Part I of this book. However, it demonstrates that safety cases in practice sometimes violate scientific rules (i.e., a violation on the regulatory level).

SLIM–Success Likelihood Index Method   The SLIM Procedure determines error probabilities based on a structured expert judgement process (Edwards et al., 1977). Experts must estimate the reliability of human actions based on a number of PSFs.

A complex action is broken down into individual tasks. Then the PSFs that can influence the successful accomplishments of the tasks are rated by experts according to their importance for the entire action sequence. The main factors distinguished are training, procedures, feedback, risk perception, and time pressure. An overall index (SLI–Success Likelihood Index) is then determined based on the estimates of the experts. The SLI is calibrated with two so-called anchor tasks for which HEP values are known. A psychological scaling method, building on the 'law of comparative judgement' according to Thurstone (1927) is used for calibration.

By using estimates of experts, SLIM attempts to circumvent the problem that data on human error behaviour are often unavailable or do not adequately fit to the situation that is to be evaluated. On the other hand, the subjective rating depends considerably on the experts who perform these estimations. Another problem is the limited number and type of PSFs used in these approaches, which are also treated as being an independent set of PSFs. This leads to the problem that the rating does not provide any proof of certainty regarding the assessment of cognitive errors or the error forcing conditions.

Methods in the Transition from 1st to 2nd Generation

A number of intermediate methods exist (i.e. methods between 1st and 2nd generation) that are difficult to classify as being members of the 1st or 2nd generation because they still persist in the thinking of the techno-morphologic information processing model but include dynamic situational factors. The value of these intermediate methods can be seen in that they showed that the 1st generation HRA methods lack essential aspects in assessing safety, namely the role of goals and intentions of the user in a working environment. Principal methods are:

- HEART (Human Error Assessment and Reduction Technique; Williams, 1988).
- INTENT (Method for estimating human error probabilities for errors of Intention; Gertman et al., 1992).

HEART and INTENT are quite similar and essentially provide an extension of the THERP tables with specific items on cognitive aspects. As an example, INTENT provides the following Human Error Probability (HEP) for the violation of a procedure and reconfiguring equipment:

- Violate procedure and reconfigure equipment: between 8.3E-2 and 5.5E-4.

The 1st generation HRA, HEART and INTENT have similar disadvantages in that the data could hardly be validated and that they still remain in techno-morphologic cognitive models, which do not fit well to the error mechanisms modelled (e.g. the violation of procedures).

Human Reliability Assessment in the 2nd Generation

Second generation HRA methods attempt to overcome the techno-morphologic modelling of cognition. They aim at considering the human as

- usually being forced to respond to more than one signal at a time (e.g. additional constraints from the environment, like weather conditions during driving or flying, management issues in the process industry etc);
- having biases, attitudes, expectations, goals and preferences (e.g. attitudes towards the technical system, goals they want to meet, like home-base syndrome in aviation).

Both issues together lead to the fact that the interrelations of dynamic situational conditions as well as cognitive aspects have to be assessed together in order to achieve a valid prediction of a possible behaviour. The most relevant methods, described in OECD (2000), Reer et al. (1999 (), or Spitzer et al. (2004), are:

- ATHEANA (A Technique for Human Error Analysis; NUREG-1624, 2000).
- CREAM (Cognitive Reliability and Error Analysis Method; Hollnagel, 1998).
- MERMOS (Méthode d'Evaluation de la Réalisations des Missions Opérateur pour la Sûreté; LeBot, 2004).
- NARA (Nuclear Action Reliability Assessment; Kirwan et al., 2004).

ATHEANA provides a complete approach for the assessment of cognitive aspects ranging from a search and identification process to a quantification component. It was developed by the US-NRC, the United States Nuclear Regulatory Commission. CREAM was a development at the OECD-Halden Reactor Project focusing on more qualitative modelling of cognitive aspects. CREAM also received several applications in the non-nuclear field. MERMOS, developed by the French nuclear utilities EDF, contains a search component and a quantification approach. It received a complete application in a probabilistic safety assessment (PSA) of the French N4 plant series. Detailed discussions of the methods can be found in Sträter (2004b).

The major disadvantage of the methods is that they are still lacking a concise data-basis and quantification approach. The lack is founded on the fact that the methods model human behaviour based on misleading triggers from the system configuration. Examples for such misleading indications are (cf. NUREG-1624, 2000):

- The system behaviour is not as expected.
- The system behaviour is difficult to interpret (ambiguous information).
- Lack of information due to failure or lack of signalization.
- The operator is forced to make decisions for or against some of his intentions.
- Inappropriate boundary conditions, like ambiguous management strategies, within the company.

Such circumstances are also strongly related to organizational and safety culture issues (see Pyy, 2001 and Sträter, 2001b). They complicate the assessment of human interventions because the task analysis as well as the quantification has to be consistent with the properties of the cognitive behaviour. As the succeeding chapters will show, the cognitive processing loop developed in Part II of this book can provide the link between task analysis, cognitive modelling and quantification.

Résumé on the Representation of Cognitive Aspects in Safety Assessments

Cognition in 1st generation HRA models   First generation HRA models for assessing human performance in safety rely heavily on the techno-morphologic processing stages. They are not appropriate for representing the dynamic situational conditions for human behaviour.

The practice of all 1st generation HRA methods is to model diagnosis processes via the available time for diagnosis as the main parameter, either by using explicit time reliability curves (TRCs as in THERP or HCR) or assuming time as the main performance shaping factor (PSF) as in the SLIM rating process for instance. By choosing time as the primary parameter of assessment, they have the advantage to consider cognition independently from the underlying model of cognition.

However, time-based approaches give a holistic view on cognition and enable only over-simplified predictions of cognition. Time is not the only parameter that influences cognition and time has various effects on cognition depending on the other parameters. One example where time is of minor importance is control of dynamic actions: in dynamic situations where time-stress would be judged to be very high, only a few errors may occur if the operator is able to perform his or her control task appropriate. Famous examples are video games that are designed to generate high time-stress and are usually used very efficiently by the user (cf. Shneiderman, 1998). In addition, event investigations show that humans can perform efficiently even if the time-windows were very short, because the actions were very straightforward and efficient. The event of Davis-Besse is an example (Reer, 1993). Operators were able to recover from a complex plant disturbance within 10 minutes.

Time hence plays a minor role for reliability. A systematic analysis of events performed by Sträter (1997/2000) showed that a more decisive factor for diagnosis reliability is the task complexity, and that time is only a leading parameter because it is confounded with high task complexity in the early phases of accidents. Experiments of Park et al. (2004) and Holy (2004) confirm that Human performance during accidents is more dependent on cognitive resolving strategies than time available. It has to be concluded that a modelling of diagnosis must not be exclusively time dependent but has to consider the cognitive strategies and processes as well. Summarizing, time reliability has a couple of principle problems from the psychological point of view:

- The TRC is only giving the time reliability for the achievement of a single diagnosis. Multiple problems (multiple failures, judgements, decisions between alternatives) are not covered.

- It does not distinguish different cognitive search processes that depend on the workload and available feedback.
- Different strategies to diagnose are not differentiated, such as for instance incomplete search for information due to confirmation biases or focusing on an isolated aspect of the situation.

Those considerations led to assessment methods, like INTENT, assessing cognitive errors by considering the particular cognitive errors (like violations) triggered by a certain error-forcing context. The methods are also lacking an underlying cognitive model explaining why a particular cognitive error results from a given error-forcing context or how different error forcing contexts lead to the same cognitive error. A typical example is the assumption that a cognitive error may be a result of either a poor design of signalization or a bad procedure layout or other deficiencies in the environment with an impact on cognitive information processing.

Cognition in 2ⁿᵈ generation HRA models Second generation HRA models give new insights through advanced cognitive modelling and by considering the individual information processing as well as the modelling of the information-processing context.

They, however, need very specific information to judge the cognitive task demands, which are often not available in safety assessments and which require a prediction of cognitive processes from a distance. The assessment relies on remote access to cognitive performance. Examples for remote settings are OECD (2000):

- The assessment of group performance in so-called low-power and shut-down states where the safety significance of complex maintenance tasks in a nuclear power plant had to be assessed and suggestions for improvements regarding collaborative cognitive performance had to be suggested (Müller-Ecker et al., 1998; Sträter, 1998).
- The prediction of errors of commission in nuclear power plants, which requires the analysis of causes for cognitive errors (Reer et al., 1999).

A cognitive model needs to take into account that the information for judging cognitive aspects is remote in order to be successful in safety cases. The scenario of failures in the system and the human reactions on the failures cannot be observed directly. Often such situations cannot even be simulated entirely if they require extensive simulation facilities. Simulators of Nuclear Power Plants for instance cannot simulate complex scenarios where operators have to perform in the control room and in the plant, because simulators only represent the control room. Therefore, the way humans' perform cognitively in such a setting cannot be simulated with sufficient validity.

The limitations and constraints of 2ⁿᵈ generation HRA led to a regression of still using 1ˢᵗ generation HRA methods, even though they had been falsified and their further use is doubtful from the scientific perspective (Hollnagel & Amalberti,

2001). Often, a limited set of HRA methods is applied in the practice of regulatory purposes. In most cases, THERP is used for HRA although it is known that it does not assess risk contributions induced by dynamic situational aspects. Even HCR was introduced for the assessment of accident management situations, although proven to produce invalid results (GRS, 1998b).

The discussion about the principal problems of modelling cognition in Part II of the book, the law of uncertainty, may lead to the conclusion that cognition cannot be considered in HRA at all. This pessimistic conclusion should not be drawn here and another possibility to consider cognition in HRA should be discussed. However, the cognitive processing loop clearly shows that the current way human error data are generated will never achieve validity because the data are collected in accordance with the techno-morphologic view and not in accordance to the generic structure of cognition and hence contains too many intervening variables in the quantitative figures.

The next section will show that this system-ergonomic framework for describing mental load and complexity suits the requirements to be generic enough while providing an access to quantification and being consistent with the cognitive processing loop.

Integrating Cognitive Performance and Safety Assessments

Classifying Cognitive Performance and Error

The link of dynamic situational conditions to the cognitive processing loop Part I of this book summarized the dynamic situational conditions that are not covered by the current approaches of including human performance into safety. It was discussed that these are most important to understand errors of commission. Figure 6.2 shows how the approach of the cognitive processing loop can resolve the issues identified.

The effect of dynamic situational conditions is reinforced by new technologies, like computerized control rooms, glass-cockpits or the distributed control of airspace and inappropriate cognitive ergonomic design will lead to critical situations or even severe accidents. While such systems have clear advantages in combining of process parameters, previews in time, explanations of system states and flexible grouping of information, they lead to more interdependent and complex interfaces, the more these features are exaggerated. The use of decision support systems will likely reinforce these demands rather than help the user. The possibilities of cognitive human failures are, in fact, increasing if the interface is not designed according to the dimensions of the cognitive control loop of the operator. The main reason for this is that the growing amount of information about the process information system means an additional task for the user with potential interference with other tasks. As an example, the operator has to remember information in computerized control rooms, which was available in parallel in conventional control rooms. A former parallel task of gathering information in the case of disturbances now turns into a sequential task. While the parallel visual-driven gathering of information requires less active cognitive resources, the operator has to search for the information actively in his brain and on the screen (remembering and combining of information).

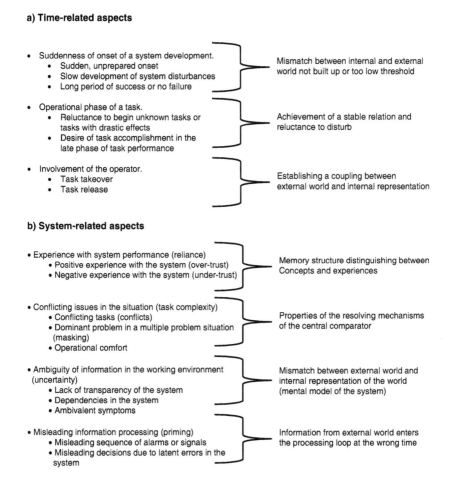

a) Time-related aspects

- Suddenness of onset of a system development.
  - Sudden, unprepared onset
  - Slow development of system disturbances
  - Long period of success or no failure

Mismatch between internal and external world not built up or too low threshold

- Operational phase of a task.
  - Reluctance to begin unknown tasks or tasks with drastic effects
  - Desire of task accomplishment in the late phase of task performance

Achievement of a stable relation and reluctance to disturb

- Involvement of the operator.
  - Task takeover
  - Task release

Establishing a coupling between external world and internal representation

b) System-related aspects

- Experience with system performance (reliance)
  - Positive experience with the system (over-trust)
  - Negative experience with the system (under-trust)

Memory structure distinguishing between Concepts and experiences

- Conflicting issues in the situation (task complexity)
  - Conflicting tasks (conflicts)
  - Dominant problem in a multiple problem situation (masking)
  - Operational comfort

Properties of the resolving mechanisms of the central comparator

- Ambiguity of information in the working environment (uncertainty)
  - Lack of transparency of the system
  - Dependencies in the system
  - Ambivalent symptoms

Mismatch between external world and internal representation of the world (mental model of the system)

- Misleading information processing (priming)
  - Misleading sequence of alarms or signals
  - Misleading decisions due to latent errors in the system

Information from external world enters the processing loop at the wrong time

Figure 6.2   Relation of dynamic situational conditions to the processing loop (related to the discussion in Part I of this book)

According to the control modes, sequential cognitive behaviour is also required for the management of accidents using emergency procedures. Interference is hence likely to occur between the dealing with accidents according to procedures and the use of the interface, because the interface needs sequential cognitive processing as well.

Particularly in serious situations like severe accidents, this interference of control modes will have a considerable influence on the reliability of human actions. In contrast with this, the currently used approaches of modelling cognition in Human Reliability Assessment are not able to describe cognitive behaviour in an appropriate framework and neither are they able to provide quantitative data for cognitive behaviour in terms of HEPs.

Abandoning the classical error taxonomy The classical techno-morphologic model of information processing understands the human being as a quite obedient, rather passive processor of information, by distinguishing processing stages like perception, decision-making and action. As discussed in the chapter on the genesis of cognitive models in Part I of this book, the distinction in processing stages results in error classifications like inadequate attention, inadequate use of memory, or inadequate decision-making.

This general picture does not change by assuming phase-based distinctions into skill-, rule- and knowledge-based levels of behaviour. The phase-based approach extends the scope of the sequential processing elements, which can fail, but does not overcome the picture of a human as a component with a specified function in the system. The phase-based approach, for instance, distinguishes mistake as related to the planning stage, lapses as memory deficiencies, or slips as execution deficiencies. It better accounts for the fact that the different stages of a sequential model are interrelated and there are no dichotic modules. However, it remains in the historical view of processing stages. This view represents, to an enormous extent, the experimental findings of the underlying experiments or investigations (i.e., self-fulfilling prophecies or an inquisitory logic regarding the structure of information processing). In general, it can be concluded that classical approaches understand the human cognitive behaviour as a techno-morphologic picture of the brain. Due to this deficiency of every techno-morphologic model of human information processing, it may be concluded that classical cognitive models are only able to focus on specific aspects of cognitive behaviour and are not able to cover the entire spectrum of cognition.

For instance, inadequate performance of the decision-making processes, like breaking rules, has to be understood as violations within the individual rather than an interrelation of a human in a contextual condition. Issues such as blame and guilt are a result of this view of human error. Other examples are design recommendations revealed to overcome human error. Design recommendations based on the classical techno-morphologic approach are more or less reduced to training, education, and better procedural guidance. They focus very little on the usability of knowledge or procedures in dynamic situations.

Design means and error prevention strategies are 1:1 related to the cognitive model used. In order to find better prevention strategies it is therefore necessary to overcome the techno-morphologic view on human performance. It was discussed extensively in Part I of this book that the sequential, techno-morphologic approach is a historical result which only survived the many attempts to overcome it because it is intuitive and makes the life of an assessor or incident investigator easy. However, the view will not improve safety. Norman (1986, p. 128) stated that that people may not even be aware of making decisions and violating rules in certain situations if they are captured in the situation. He also stresses that events that were experienced in former times are the basis for many biases, even if they are not consciously available at the moment of the decision-making act. Accusing a human who made an error as being a rule breaker is a convenient but wrong treatment by an investigator. Cognitive errors cannot be sufficiently described in using macro-level models like stage-based approaches.

Omissions and commissions The overall risk of hazardous technologies is usually mis-assessed regarding human contributions, if the wrong cognitive model is applied. The deficiencies of classical error taxonomies result in a critical issue because they lead to safety assessments overlooking critical contributions to the overall risk. Chernobyl is such an example of the relevance of cognition as a risk contributor. Overconfidence of the crew in its own competencies was never considered as a contributor in risk assessment, but was one of the contributors (among others) to the accident: the operating crew of Chernobyl was the most effective one until the accident happened. Similar observations can be made in any other technology as well. Theis (2002) could find a mis-evaluation by the factor of 200 for simple routine tasks during car driving. Cartmale (2004) found a factor of 100 for more complex air-traffic-control tasks between the assessed risk in safety case and the actual observed risk.

In order to have a realistic safety assessment, the active human contributions to risk have to be included throughout all phases of the system life-cycle (e.g. design, construction, management and operation). An important distinction in safety assessments in this respect is the one between omission and commission errors. An omission may be defined as the lack of performing a certain task or mission. A commission error may be defined as an additional failure introduced by a human action into the system.

For omissions, we can observe a many to one relation: many different elements of an entire system can fail with a certain probability to omit their required function. Many-to-one-relations can be analysed and predicted easily, because the outcome is defined.

Any commission has the potential to change the course of accidents, either a not-required technical behaviour or a not-required human action (e.g. the intervention of the shift in Chernobyl initiated the bad development of the accident; without their intervention, the accident would not have happened).

Compared with omissions, where the problem space is precisely defined, the problem space of commissions is ill-defined. The prediction and prevention of commissions is therefore much more difficult. There is hence a one-to-many-relation between a commission and the possible outcomes of it. These relations are of dynamic nature and strongly context dependent. The cognitive control loop allows dealing with both properties. The method therefore can be used to avoid heavy discussions, which usually happen between analysts of the technical system and analysts of the human part. These discussions often lead to inappropriate assessments and unsuitable constructions of the safety assessment model (e.g. worst-case modelling). The assessments often try to cover the human impact on risk quantitatively without having any feeling or proof whether the modelling or the assessment is really covering all human influences.

Cognitive errors as a result of the cognitive processing loop Some authors suggest avoiding talking about human errors at all as one way out of the dilemma (e.g. Fujita & Hollnagel, 2004; LeBot, 2004, Amalberti, 2001), because the error is describing the external performance (task performance) of a human within a system rather than the mechanisms behind it. An error of commission (EOC) may be reasonable from the operators' point of view and he or she may be convinced to do the correct thing but may fail due to technical circumstances, like incomplete or misleading information about the technical system states.

Indeed, it is difficult to speak about errors for normal operation of the cognitive system. The term *Human Error* has, to some extent, similarities to the field of software reliability. Software does not fail in the way that mechanical components fail (software does not 'wear out') but it does exhibit erroneous behaviour under certain contextual conditions due to internal latent inadequacies (Shorrock & Sträter, 2004). Software fails by wrong demands or demands out of the specification of the software, i.e. under conditions that are outside the capabilities of the software. The same holds for human errors. They are a result of an inadequate relation between the internal and external world. Part II of the book encountered a couple of revisions of the techno-morphologic view, in order to come up with an enhanced view on human error and performance. The main findings beyond the techno-morphologic were:

- There is no distinction between long- and short-term memory (memory span).
- Cognitive processes are a result of dynamic binding of the internal and external world.
- Human cognition and behaviour depends on goals and situational conditions.
- The memory span is a result to the dynamics of a situation.
- Reaction time is a result of the memory span.

A number accidents and events were discussed in Parts I and II, showing the importance of these aspects for human reliability. They cannot be explained in the classical view of cognition. A couple of statistical proofs throughout a variety of industries, from the aviation, nuclear and automobile industries will be provided in the course of this part.

The processing loop suggests abandoning the techno-morphologic error classification and introducing a classification according to the generic structure of the processing loop, as outlined in Figure 4.10. The error modes as represented in Table 6.1 can be distinguished based on the cognitive processing loop.

A Descriptive Decision Model for Safety Assessments based on the Processing Loop

For considering cognition in safety assessments, the cognitive processing loop has to be linked to observable information in the technical system (e.g. the controls to be manipulated) and the organization of the system. Figure 1.2 (in Part I) provides a hint of how that this link can be achieved using the concept of constraints. The influences of decisions by the management, procedures, interface, communication or tasks on safety can be represented in safety assessments if they are understood as constraints for the person who has to decide and acts at the sharp end. The cognitive processing loop is the common denominator where all these influences come together.

Table 6.1   Cognitive error taxonomy based on the cognitive processing loop

Aspect of the cognitive processing loop	Likely behavioural error (example)
Experiences or concepts not matching the external demands	
Concepts not established	Omission
Experiences made in different contexts	Commission, violation
Resolving mechanism in the case of mismatch between internal and external demand (plus emotional reaction)	
Fixation on established coupling / actual cognitive relation	No adaptive behaviour, sluggishness, too late reaction on contradicting information
Information ignorance / utility-oriented processing	Eagerness to act, hastiness
Expectancy-driven / habitual or experience-driven processing	Neglecting side-effects, frequency oriented reasoning
Information and goal elaboration / rationale processing	Reluctance, rather slow but thorough reaction on external information
Adding information from the external world / enrichment	
Information provided matches the internal demands	Search for information
Information provided matches other than the desired concepts and experiences	Commission

The Titanic is a classical example involving design-, construction-, management- and operational-errors that ended in the decision of the captain to choose the risky North-Atlantic route instead of a less risky southern route.

In order to integrate cognition into safety assessments, safety assessments link the system to be assessed with the human behaviour and the influences on it. This requires taxonomies of cognitive tasks in the first place. The taxonomies should be able to describe the given situation for the operator in such a way that the task description is ideally based on external, objective information. However, it also needs to represent the cognitive activities that the operator is forced to make. The control modes described in the system ergonomic framework fulfil this requirement. It secondly requires taxonomies of cognitive errors. This taxonomy should be able to link the cognitive model with the likely behavioural result. The table above, describing the link of the resolving mechanisms and the cognitive processing loop, provides this taxonomy. Finally, the cognitive processing loop is able to link all the different constraints in cognitive processes and herewith links causal aspects (genotypes) with behavioural aspects (phenotypes). The binding elements of the cognitive processing loop are the control modes and the cognitive resolving mechanisms. The overall chain is then:

- Constraints → Control modes → Resolving mechanisms → Behaviour

Several different names are provided to the constraints by various models. Helmreich (2002) described constraints as threats in his Threat-Error-Management Model (TEMM), the HRA still uses the term PSF (Performance Shaping Factor) while some use the term error-forcing context (EFC). Control modes describe the effort required by the human to bring all the different constraints into a homogeneous and consistent picture (a stable relation). Montmollin (1992) calls the resolving act 'signification'. In cybernetic terms, the cognitive system has to accomplish a recoding effort. In the sense of the stress/strain concept of ergonomics, control modes describe the strain of the human in a given situation based on the external threats. Resolving mechanisms of the processing loop are needed to minimize the recoding effort. Herewith the processing loop allows describing the link between basic cognitive processes and error types.

From constraints to control modes For integrating cognitive aspects into safety, the term 'task' needs to get a more realistic meaning. Rather than understanding the task as the isolated task a user has to fulfil given his job-description, a task in the cognitive sense contains a number of additional aspects. Figure 6.3 summarizes the main spheres of tasks that might affect the cognitive processes of a person.

The cognitive task is to be understood as a composite of the task in the technical working environment (the so-called Man-Machine System), the external threats from management, external organization (like stakeholders and regulators) and society (including the political level). The cognitive task eventually also includes personnel knowledge and experience (as gained during education and work practice) as well as personal goals (e.g. life objectives). An important final contribution to the cognition stems from the ad-hoc dynamics in the situation. During the mid-air collision at Lake Constance, for instance, the ad hoc condition was that the second controller went for a break and left his colleague alone; coincidently, the same happened on board the Russian plane, where the pilot left the co-pilot alone for a comfort break.

As a result, the human performance may have, in the worst case, a completely different outcome than the desired actual task needed in the working environment. Occupational safety has considerable experience with human behaviour not related to the actual task demands. For instance, users take safety systems out of operation to be more effective. The reason is that they are paid on a piecework basis. A request to comply with safety rules (like two hand operations at a punching machine) usually contradicts productivity and personal income, which diminishes the compliance with safety regulations. Shull (2002) investigated the reasons for rule breaking within the group and found amongst others:

- Unnecessary, unrealistic or complicated procedures and rules.

- Bad working environment (ergonomics, resources).
- Lack of attitudes (values / beliefs) towards rules.
- Counter experience using rules.
- Personal advantages.
- Inadequate situations (situations where rules are not valid).

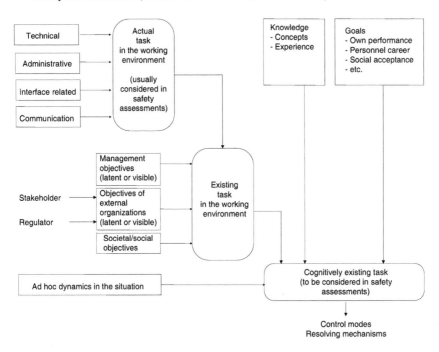

Figure 6.3   From simple tasks to cognitively existing tasks

From control to cognitive resolving mechanisms The system ergonomic approach presented in the previous section showed a way to link task descriptions with the cognitive control modes. In the understanding of the stress/strain concept of ergonomics, the control modes are needed for minimizing the strain of the operator. Any complex task may contain a set of cognitive coupling modes and may lead to several demands of the cognitive processing loop. Hence, virtually all elements of the processing loop may lead to a cognitive error in real settings. However, for prediction it is useful to break complex tasks down to the coupling modes in order to find potential cognitive demands and sources for cognitive errors. Depending on the relationship between the internal and external worlds, several control modes were identified. Applying the generic error classification to the control modes results into Table 6.2. By classifying a task according to the control loops, the table provides the most probable cognitive errors. The table uses the cognitive control modes as introduced in Table 5.1 and applies the error classification to the control modes.

The cognitive errors result from the information-processing loop. The table summarizes the results of several authors who investigated the relation of cognitive errors to control modes, like the contextual control model of Hollnagel (1998) for the nuclear industry, the investigation of nuclear incidents by Sträter (1997/2000), the experimental investigation on car drivers' behaviour by Theis (2002), and the industrial maintenance investigation by Köhler (2001).

Table 6.2   Parts of the cognitive control loop likely involved in cognitive coupling

Cognitive coupling		Main elements of the processing loop involved			Examples
Relation (Internal - external)	Control mode	Experiences or concepts	Resolving mechanism	Information enrichment	
- ↔ -	Interact		■		Human machine interaction
M | -	Reflect	■			Problem solving, thinking of several alternatives
1 ← 1	Monitor		▨	▨	Watching information like waiting for the green light at cross light
1 → 1	Perform	▨	▨		Switching a valve, piecework, one task per time
1 ↔ 1	Track	▨	■	▨	Regulating the altitude of an aircraft or the flow of a pipe
1 → 1	Operate	▨	▨		Reading a label or text, use checklists; situational conditions change
M | N	Imagine	■		■	Imagination of process behaviour (e.g. shut-down, start-up)
1 | 1	Expect	▨		▨	Combination of distinct information with semantic meaning
M → N	Coordinate		▨	▨	Task is linked to different other tasks
M → 1	Follow		▨		Task is a special and clearly defined task
1 ← N	Identify		▨	■	Clear hint on information and one solution
M ← N	Recognize	▨	▨		Unspecified information and no specific solution
1 ↔ 1	Align	▨	■	▨	Multiple connections of information, task not structured
M ↔ N	Compare	■	■	■	Task is structured or hierarchically ordered

The mapping of the coupling is also logically inferable from the cognitive coupling. The column 'relation' reflects the relation between the internal and external worlds according to the discussion of the system-ergonomic approach. The column uses the database notation: 'M' and 'N' means there are several independent sets of information in either the internal or the external world. '1' means there is one coherent set of information. '-' means there is no set of information present but it will be generated based on the direction of the implication. The sign ' | ' means, there is no coupling between internal and external world required to perform the control mode. The sign '↔' means there is coupling in both directions, '→' means there is an implication from the internal to external world, '←' means there is an implication from the external to internal world.

Successful imagination for instance mainly demands the internal world (the memory structure) and the information available in the external world (information enrichment). It is less related to the resolving mechanisms. The same holds for expectations. However, as soon as the internal world interacts with the external world the resolving mechanisms come into play.

The shaded parts of the table indicate the link. The higher the demand for interaction, the more the resolving mechanisms come into play for proper cognitive processing (dark shaded fields in the table). For simpler tasks, the demands are less (light shaded parts of the table).

From cognitive resolving mechanisms to behaviour The next step is the link of cognitive resolving mechanisms to behaviour. Resolving mechanisms are used in case constraints cause cognitive dissonance. The eventual action is a result of solving the trade-off by achieving cognitive consonance between the situational constraints and the internal goals.

The resolving mechanisms are only needed in the case of existing dissonance between the internal and external worlds. They are described in detail in Part II of this book and are equivalent to the error classification outlined in Table 6.1 (above). Four internal resolving mechanisms have been identified along with the effect of gathering additional information from the outside world in order to reduce uncertainty about the external world.

It was discussed that the resolving mechanisms are basically a reduction of the threshold of the central comparator. This process is an iterative process beginning with a small reduction, which is continuously reduced until a solution is found. Figure 6.4 unfolds the cognitive processing loop into a two-dimensional paper-based decision process. Although limited in the explanatory richness of the model of the processing loop it provides the essential aspects discussed in Part II of this book and outlines how the resolving mechanisms can be used for the prediction of cognitive behaviour. The figure includes the experiences gained with event investigation of humans in critical events as well as simulator experiments revealing that time efficiency goes before spatial optimization (related studies are discussed in detail in Sträter, 1997/2000; Linsenmaier, 2005; Theis, 2002; Schweigert, 2003).

The resolving mechanism of changing the threshold of the central comparator is modelled in this figure according to a distinction provided by Rasmussen (1982, p. 18). The levels of abstraction are compiled in an engineering sense distinguishing five levels of abstraction where each represents an iterative processing cycle with reduced threshold. In case none of the resolving mechanisms is successful, operators may start to abandon the initial goal. Giving up goals is a difficult process and experience shows that it usually happens very late in the development of accidents. For instance, evacuation of persons in the surrounding area is instructed. Therefore, a goal change is not represented in the figure.

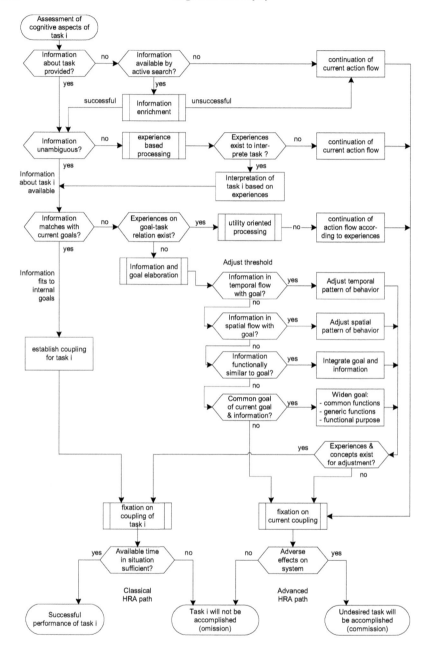

Figure 6.4   The unfolded cognitive processing loop from resolving mechanism to behaviour

The resolving mechanisms distinguished by Rasmussen (1982, p. 18) and represented in the figure are:

- Physical shape: In the case of mismatch, control elements are used with a similar physical appearance according to the Gestalt laws (e.g. same shape, spatial proximity). For example, in the case of waiting for the elevator to go up many people waiting tend to press the down button with the hope the elevator comes faster.

- Physical function: In the case of unsuitable elements with similar physical shape, controls are used which have the same function (electrical, mechanical or chemical). For example, in the case of failure of one control element, operators might tend to use the control elements on the same control panel with similar shape.

- Common functions: In the case of unsuitable elements with similar physical function, systems are used with the same general function. For example, in the case of failure of the brakes during car driving, drivers may choose the redundant hand brake.

- Generic functions: In the case of unsuitable elements with a common function, the causal structure of the system is investigated according to comparable means to perform the task. For example, in the case of failure of the heat removal systems in a process control due to blocked valves, operators may start to think of how to bypass the blocked pipes.

- Functional purpose: In the case of unsuitable elements with a generic function, general interrelations of the goals to be achieved are considered. For example, in the case of failure of heat removal systems in a process control due to blocked valves and no bypass opportunity, operators may start to launch accident management measures like cooling the system with the help of the fire brigade.

Considerations on the Possibility to Quantify Cognitive Aspects

Quantitative assessment needs stable structures to build the statistics on. Prediction of cognitive behaviour would be relatively easy if people do what they should do from the system designers' point of view and if system designers' could foresee all possible problems a designed technical system may be exposed to in the future. In this case, we would have a clearly specified environment, the action is prescribed and the system behaviour determined.

This is certainly not the case. The system designer can make some errors in design and constructions before a technical system gets into operation. These may, accompanied by additional constraints of the management, lead to human decisions during operation, which are not adequate in a given situation.

First generation HRA did a quite lousy job on establishing well-validated data. Data were often generated by expert judgement and these data persisted as well as the techno-morphologic processing model behind them. The many problems of the techno-morphologic information-processing model also had an impact on the possibility to validate the data. Validation was not possible because the cognitive model behind it did not allow it to come up with stable statistics of human error rate. The model was inherently unstable due to the limitations of techno-morphologic roots. As stated above, the data are however still used in safety assessments although they would not survive any critical look at the statistical standards on quantitative figures. On the other hand, the trust in new approaches has not yet been obtained.

Problems of Quantification

An essential problem of HRA methods is the valid quantification of human behaviour. In current Human Reliability Assessment (HRA) practice, one sometimes observes many arbitrary quantitative figures of the Human Error Probability (HEP) due to the problem that no systematic validation or even consideration of the numerical bases was performed. Many discussions usually come up in regulatory reviews or scientific conferences about the validity of the data used in an assessment. Indeed, the assessments often reflect the discussion that went on between regulator and licensee during a Probabilistic Safety Assessment (PSA) rather than a profound database that both can use as common ground for their discussions and decisions. This data problem may raise the following issues of PSA/HRA:

- HRA is not seen as a credible contribution to PSA, although some other areas of PSA have similar problems (e.g. some common cause problems, fire, and software). Even the entire credibility of PSA or safety cases may suffer, if the human part is not considered adequately.
- The importance of PSA and HRA as a tool for decision-making and safety assessment and evaluation is questionable in some cases, although Risk Informed Decision-making (RID) is one of the most powerful approaches to tackle human issues in regulatory tasks (Sträter, 2001b).
- New problems in HRA, like the Errors of Commission (EOC) issue, are not tackled since the data problem has not been resolved.

Better data for HRA would be a decisive input to enhance HRA and PSA as well as to enhance the use of these tools for systematic prediction and avoidance of errors. The data issue is apparently not an easy one: Although they have received at least some validation studies, the existing data-tables of the THERP handbook, for instance, are still speculative. The only reason why the data of the THERP method survived the criticism is the trust the method gained in its years of application within the user community. Other methods often lack a traceable way of how the data are generated. Additional problems of using of data are due to differences in

- the cognitive, technological and organizational model that is used,
- the type and extent of errors and PSFs that are seen as significant,

- the unavoidable uncertainties due to the limitations of the sources of data to be used for assessment (experiments, observations or operational experience), and
- how single observations are comprised into an overall assessment in a mathematical sense.

The solution of the data-problem seems to be difficult to achieve since it includes psychological, mathematical and pragmatic constraints of HRA. Most difficult in this framework is that the experts from the different fields are sometimes convinced that only their view on the problem is the most important one. If a good and serious quantification should be achieved all these perspectives have to be worked out in close cooperation.

Psychological Soundness of Quantitative Figures

Psychological scaling theory has been developed for many decades to overcome the problem of generating quantitative figures for psychological issues. The most popular examples are known from intelligence testing, manpower planning and personnel development. The methodological approaches used there also hold for the problems of Human Reliability Assessment. The approaches are known as scaling problems. In order to establish a coherency of psychological soundness and quantification, scaling starts with distinguishing the data levels of quantification.

Data levels In judging the various methods, one must consider the quality of the data that these methods used for quantification. Various possible procedures involve estimates by experts, experiments, simulator studies, or an empirical acquisition of data from practical operational experience.

Procedures involving estimates by experts constitute the simplest possibility for obtaining data on human reliability. In the case of simulator studies, one performs operational runs within a simulator, which needs specification of the effect in advance, and one qualitatively or quantitatively analyses and evaluates the behaviour of the operators. Experiments supply data with standardized and defined marginal conditions. Looking at practical operational experience, finally, realistic data material is available in the form of events.

Based on the level of the data as such, the advantages and disadvantages of the various data sources can be discussed. In general, we differentiate as outlined in Figure 6.5 (see e.g. Coombs, 1964).

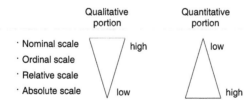

Figure 6.5   Hierarchy of data levels of measurement

The scales outlined in the figure define a hierarchy of the levels. Any quantification has to start on the nominal scale level, no matter whether it is applicable to a psychological topic or technical topic. The next highest level is the ordinal scale, followed by the relative scale and finally by the absolute scale.

Absolute scale means the data would allow supplying HEP values within limits of 0 to 1. Methods working with relative scales can only provide statements about the interval (e.g. 'if the frequency of error 1 is 10 and the frequency of error 2 is 20 then error 1 is twice as important as error 2'). Methods involving ordinal scales can only set up priorities, such as, for example, event 1 is more likely than event 2.

The nominal scale finally provides what is called a *Construct* in psychology, namely a basic assumption that a certain psychological aspect is based on a certain property of the human mind (e.g. intelligence is such a construct, where we assume that there is something like intelligence present in our minds). This construct is sometimes called *Latent Trait* because it is usually not directly observable. The construct may have an explicit part and an implicit part. Intelligence, for instance, is the explicit part of a construct; the implicit part of the construct is that intelligence is a property of humans (and not stones or something else). Usually the implicit part of a construct is a forgotten part in the quantification of human errors. The THERP-item 'Omitting a step in a Procedure' for instance has the implicit construct that there are operators with a certain qualification omitting something.

The way to generate and validate data has to consider this hierarchy of data-levels. Before we come up with a number for a certain construct, we should validate whether the construct itself makes sense. For instance, the construct 'Adherence to procedures' would imply that there is a certain ability of humans to adhere to procedures that is more or less an inherent property of all humans.

The ordinal scaling needs to be proven after the construct is validated. This includes a check whether the chosen construct leads to transitive or intransitive statements. A transitive statement would be, for instance:

- If person P1 adheres more often to procedures than person P2 in situation S1 then P1 also adheres more often in Situation S2. If this is not the case, we observe an intransitive statement about the construct 'adherence to procedures'.

Thus, an intransitive statement is always a hint for wrong assumptions on the construct itself. Intransitive statements can be resolved if the construct we assumed as valid for the observation is not valid but can be substituted by another one. If we have an intransitive statement (person P1 adheres more often to procedures than person P2 in situation S1 but P1 adheres less often than P2 in situation S2), this may be resolved if the different situational conditions can be represented by a construct (e.g. by the construct 'time pressure'). If this is the case, we may start a scaling of the construct 'time pressure' instead of the construct 'adherence to procedures'.

On the next level, the relative scale level, we have to ask ourselves about the following:

• If it is observed that P1 adheres in twice as much events to procedures than P2, is he then in general more adherent by the factor of 2, i.e. is the observation indeed a representation on the mathematical level and can be expressed by a numeric factor?

At this point, HRA methods usually introduce anchor points, reference points, or perform a systematic calibration since it is virtually never achieved to get a representation of the entire scope of a scale.

Finally, the basic question of the relative scale would be whether all persons might be arranged according to an absolute scale ranging e.g. from 0 to 1 as required if we make a HEP assessment. The arrangement therefore has to be represented in a certain distribution of a one-peak density distribution with a monotonously increasing cumulative density function.

The view on the distribution   The different data levels can be easily seen in any HEP distribution. The X-axis usually represents an operationalized construct that is to be quantified. The relation of different points on the X-axis represents the ordinal scale (e.g. P1<P2). The scale on the x-axis represents the relative scale and finally the area under the density function reflects the absolute scale. Figure 6.6 expresses these relationships. Note that human errors are log-normally distributed.

The rationale behind assuming a lognormal distribution is the fact that the general underlying construct for a human error is 'human ability to perform a certain task correctly'. As users of a system are usually trained and skilled on the tasks to performed, the distribution is slightly shifted to the left. The dimension then goes from rather skill-based on the left to rather knowledge-based on the right. The shape of the curve, an assumption introduced by Swain & Guttmann (1983), was validated several times. Sträter (1997/2000) could show that it holds for the reliability of nuclear operators. Häusler (2005) showed it for pilots' team capabilities. Schweigert (2003) could show it for car-drivers. The log normal distribution could be validated in both the cognitive levels and the time users spend on system monitoring or understanding. This is another proof of the fact that decision-making, memory span and reaction times go hand in hand, as described in Part II of this book.

This means that the density-function can be fully interpreted using the processing loop. The function then shows the number of persons who need a certain number of processing loops ranging from 0 iterations to X iterations of the cognitive cycle.

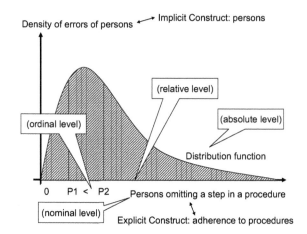

Figure 6.6   The view on the distribution

A distribution that can represent all observations in a homogeneous way means that the distribution has one peak and not several and it is increasing in a monotone way if the distribution is drawn in a cumulative fashion. For any quantification, it is important to judge the quality of the data regarding:

- The coherence of the construct with the nominal level of distribution (e.g. is 'omitting a step in the procedure' a sufficient nominal expression of the construct 'adherence to procedures' or is something missing like 'violating a procedure' for instance).
- The possibility to order the nominal categories in a uni-dimensional order that reflects a certain relative meaning or level (e.g. omitting a step in situation S1 is less critical that in situation S2).
- The possibility to count certain items related to the relative level (e.g. number of events or persons observed).

The Law of Intransitive Statements

The old debate of the possibilities of quantification of human performance can easily be decided based on the idea on transitive vs. intransitive statements.

- A statement is transitive if the three observations fit the following relation: *a>b; b>c and a>c.*
- A statement is intransitive if the three observations fit the following relation: *a>b; b>c and a<c.*

Is quantification possible or not? Based on these considerations, an old debate can be reconsidered: is quantification of human error possible or not? The answer to this question depends on the coherency of the data levels.

This question can clearly be answered with 'yes' if there are only transitive statements on all data levels of the distribution.

For instance, a person (or group of persons) P1 performs in all situations with fewer errors than a person P2 and person P2 has fewer errors than person P3. In this case, the law on transitivity is fulfilled: P1<P2; P2<P3 and P1<P3.

Slight intransitives in the distribution can still lead to a 'yes' if the same or a similar distribution as Figure 6.7 can represent them. Here the statement depends on the situation one is considering: P1|S1<P2|S1 and P1|S2<P2|S2 but P1|S2>P2|S1. Such slight intransitive statements can be resolved without changing the underlying construct but by introducing situational factors in the quantification. Using weights for PSFs while keeping the general construct of 'adherence to procedures' is one example for a resolvable intransitive statement.

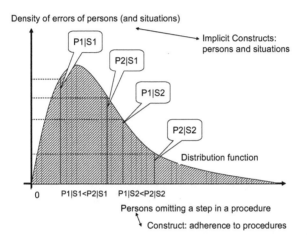

Figure 6.7   The view on the distribution with resolvable intransitive statements

In Figure 6.8 there is an intransitive relationship of person P1 and person P2 in the two different situations S1 and S2. This intransitive relationship, however, can be compensated if both can still be represented according to the shape of one density distribution, so that the density distribution is capable of compensating the intransitive relationship. The density distribution reflects in this case one explicit construct 'adherence to procedures' and two implicit constructs: 'persons' and 'situational conditions'.

Such quantification is to be considered carefully since it does have the potential of leading to uncertainties in prediction. Since this is, on the other hand, always the case in practice, a permanent monitoring of the information that feeds the distribution (e.g. operational experience, experimental data) has to take place.

The question can clearly be answered with 'no' if there are intransitive statements leading to a violation of the distribution as the figure presents.

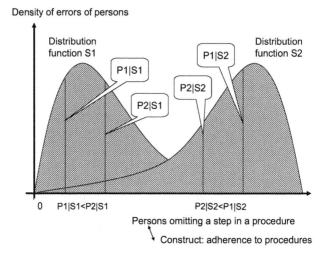

Figure 6.8   The view on the distribution with non-resolvable intransitive statements

In the investigation of Sträter (1997/2000), such a non-resolvable intransitive statement was observed. The adherence to procedures was observed to be completely different for the following two cases:

- Before an initiating event, the adherence was low.
- After an initiating event, the adherence was high.

From the mathematical point of view, one can calculate an average and a HEP value across both distributions. However, this is wrong from the psychological point of view. The psychological point would propose, in this case, separating the two situations quantitatively and making quantitative predictions in terms of HEPs separately for the construct 'adherence of procedures | before an initiating event' and for the construct 'adherence of procedures | after an initiating event'. Another possibility would be to define and test a new construct 'event type' for future quantification.

Summarizing, the hierarchy of data levels suggests that comparisons of data from various sources are possible. The approach also does not suggest that qualitative data be of a different nature than quantitative, but that they represent a necessary and reasonable continuum. This continuum of quantification has

- An underlying construct.
- An operational definition of the construct (nominal level).
- An error dimension of the construct (ordinal level).
- A distribution for the construct (relative level).
- A one peak, monotone, intransitive shape (absolute level).

Data fulfilling these five characteristics are called *Stable Distributions* hereafter.

Quantitative Assessment of Cognitive Processes

Data level and HRA   Each HRA method must supply data on the absolute scale level in order to be able to be employed in a PSA. Upon closer examination of the various methods, however, we find that the existing methods accomplish this only very rarely.

Experiments and simulator studies represent the only possibility of supplying data on the relative scale level and thus providing at least a statistical estimation value for data on the absolute scale level. However, we encounter the problem of the transposability of the probabilities obtained to the situation that is to be judged if such data are used. The THERP method is, according to Swain & Guttmann (1983), at least partly based on experimental data; therefore it represents to an (albeit unknown) extent, a method on the absolute scale level.

Methods involving estimates by experts typically provide data on the ordinal scale level. These methods employ scaling procedures in order to 'raise' the data level. However, both the mathematical setup of the scaling method and its underlying prerequisites must be plausible and must be capable of being tested. A typical example is SLIM, which generates estimates on the absolute scale level from data on the ordinal scale level using the absolute scale level of the anchor tasks for the calibration of ordinal scale estimates by experts for the individual tasks.

Regardless of whether it is holistic or decompositional or whether it is oriented by PSF or error types, each HRA method needs data to determine the probability of a human error; therefore, the underlying data are exposed to particular criticism. This criticism is the more intensive, the lower the data level of the method. Data are often insufficiently accurate or they are not realistic enough. This uncertainty in data is further increased by the fact that HRA methods so far are hardly being subject to a systematic investigation of the validity.

In addition to the inaccuracies in available data, the methods likewise are not in a position to supply data for more recent problem fields. For example, the incompleteness of the methods (for instance, in the area of the organizational effects or cognitive errors) can lead to a mis-estimation of the total risk. Cognitive errors are given only insufficient consideration but are important, particularly for estimating human reliability in future technical systems (for example, computerized control rooms) and, along with organizational aspects, they are significant in terms of internal system measures (AM-Accident Management).

The data issue constitutes a main problem in the HRA methods; this is why attention must be focused on this problem. Both the validation of existing data sources and the evaluation of cognitive and organizational errors should be possible with the help of the data derived from practical operational experience. A method for analysing data derived from practical operational experience, furthermore, must be flexible enough in order to cover different ways of quantification (error related, PSF related, or time related).

The need of stable distributions Quantification does only make sense if the statistical data have a stable distribution. The limitations of the techno-morphologic information-processing model have already been stated several times. These limitations are also transposed into the quantification of human error. Therefore, it is well justified to be very reluctant about the quantification of human errors if the wrong underlying cognitive model is used. The quantitative figures of the assessment method HCR, for instance, could not be validated because the model behind was the techno-morphologic processing model plus the misuse of the ladder-model by assuming the cognitive levels as task-specific rather than situation-specific. Reservations against human error quantification are therefore often raised (e.g. Dougherty, 1992; Adams, 1982; Fujita & Hollnagel, 2004; Amalberti, 2001). However, these reservations assume an inappropriate cognitive model behind quantification. According to the discussions in the previous section, the inappropriate model is equivalent to an inadequate underlying construct, which leads to a wrong operational definition of the construct and multi-peak transitive distributions.

A quantification of human error can only be successful if it considers the architecture of the cognitive processing loop. It was discussed that events are an ideal source to understand human experiences as part of the processing loop. The next section will show that they are also suitable to generate quantitative information about human reliability.

Therefore, a more realistic approach of human information processing can be achieved by exploiting operational events. The OECD Working Group Risk for instance is currently undertaking a task on standardizing event collection in order to make use of the data for Human Reliability Assessment (HRA) across the OECD member countries. An initiative of the VDI (Society of German Engineers) is working on basic requirements for event reporting regarding human error (Bubb et al., 2004).

Approach of Using Events for Quantification

Events allow reflecting the dependencies between causes for human errors and providing herewith evidence for statistical soundness of a quantification approach. Even if event data might be seen as insufficient for generating quantitative statements in terms of human error probabilities, as sometimes stated, they can still be used for proofing the underlying construct of human behaviour or for proofing whether a distribution has transitive statements. Figure 6.9 provides the approach of generating statistical data based on the above-mentioned CAHR (Connectionism Assessment of Human Reliability) method. It was developed in the nuclear industry and was applied to aviation as well as maritime and occupational safety events (Sträter & Bubb, 2003; Sträter, 2004a). To build up the experience-based knowledge base of the model, 237 operational events with 165 sub-events were analysed regarding human performance and related contextual conditions (Sträter, 1997/2000; Sträter & Bubb, 1998). The knowledge base of the method is under continuous development (see IALA, 2001b, Linsenmaier & Sträter, 2000).

Experience shows that a collection of about 50 events provides already a quite detailed picture of the interrelations of the various causal factors (Sträter & Reer, 1999).

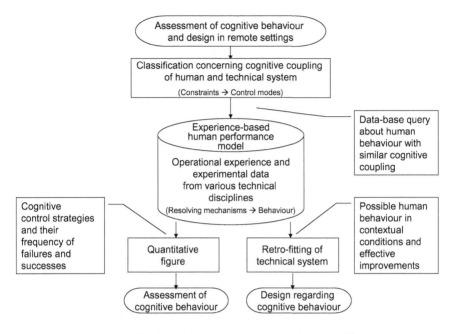

Figure 6.9 **Approach for using retrospective incident data for prospective design and assessment**

The approach suits the complexity of human information processing by evaluating the dependencies between human behaviour and different causes. In relation to the scaling requirements, the approach enables us to:

- Define the nominal level with a certain taxonomy.
- Define the error dimensions and causes.
- Define the frequency of errors under certain circumstances.
- Enable ranking of errors under certain circumstances.

As shown in the next section, the predictive approach was successfully applied in the nuclear industry, aviation, automobile industry and occupational health for predicting human reliability and for finding appropriate designs regarding cognitive aspects.

In the long term, the approach can serve as an integrated human error management approach that consists of incident analysis methods, and design decision support systems. Such a management system needs continuous feedback of all working levels (from working level to organizational and regulatory level).

Human error management therefore must extend incident analysis to other than the working level and must exploit incident data much better. The current situation of filing incidents into databases, which cannot be evaluated according to the needs of designers or managerial decision makers, needs to stop.

The approach presented in the following for prediction is limited by the data but not by the process. As current event evaluation methods do not fully address all operational levels, the event evaluation methods need to be enhanced in this direction. An event-based predictive method needs to support the assessment of those operational levels that are investigated with the event analysis method. The database also needs to represent the complex interrelations of the levels affected (operation, maintenance, design, organization and regulation).

Incident reporting has to be improved regarding coverage of all operational levels (Van Damme, 1998; IAEA, 2001). In order to assure a continuous loop from incident analysis to design and safety assessment, the parties involved in these activities need to enhance their common understanding. Therefore, the framework also needs accompanying processes to facilitate and moderate the potential interests and constraints of parties dealing with error mitigation.

An argument against using events for prediction is often that the behaviour observed in daily events does not reflect behaviour in emergency-situations, which needs to be predicted. This argument is tainted with the techno-morphologic processing model as a basis of the event analysis method. An event analysis founded on the cognitive processing loop overcomes this restriction. This can be demonstrated by a couple of validation studies described more detailed in the next chapter. It can be shown, for instance, that the behaviour of nuclear operators can be exploited to predict car drivers' behaviour, if the assessment exploits the cognitive processing loop as the underlying cognitive model. The experience may conclude to use operational events, in combination with an advanced model of cognitive processing, to predict human behaviour in accident situations as well as managerial or organizational tasks.

A Calculus for Quantification

As the cognitive processing loop is a result of the coupling of the external and the internal world, the probability of failure in cognition must also be a mathematical relation between statistical expressions related to the external and the internal worlds. Two possibilities exist. The first possibility is that the external situation is specified exactly and a representative set of human actions were observed in exactly this situation. In such cases a direct expression of errors n per opportunity N may be calculated according to the classical definition of human error probabilities $HEP = n/N$.

The second possibility results from the fact that quantification of cognitive error means not predicting individual performance but expected errors based on the observations of a range of persons having different capabilities and constraints during the task to assess. Even the behaviour of one person may differ in the same situation because of different levels of vigilance for instance. In addition, the conditions of the external world have uncertainties. For instance, the probability for shift handover in an air-traffic-control centre should be assessed. Sometimes a shift handover is performed under more time stress, sometimes under less. Sometimes the traffic is dense sometimes not. Overall, both the internal and the external worlds have inherent uncertainties.

A cognitive error results if the dissonance between the internal and external worlds cannot be resolved or is resolved with a well intended but unsuitable decision. If the human is not completely decoupled from the external world, the internal and external worlds always have some overlap. This overlap represents the potential for errors of omission and commission.

The behavioural outcome of the cognitive behaviour can hence be understood in a similar manner to signal detection. If the external world requests a particular behaviour but the internal world does not respond, we usually speak about an error of omission. If the internal world provides a behaviour that was not requested by the external world (e.g. due to mental utility), we speak about an error of commission. Both internal and external worlds have uncertainties with respect to a certain situation to be assessed. Figure 6.10 represent how quantification then results from these considerations.

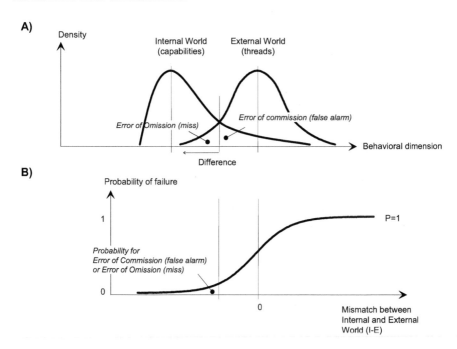

Figure 6.10 Assessment of cognitive errors in uncertainties of the internal and external worlds

Overall, the probability for either an error of omission or an error may be expressed by the probability of missing a demand from the external world and an error of commission may be expressed by a false reaction of the internal world.

Both are reflected by the area left, respectively right, of the crossing point of both uncertainty distributions. The crossing point as represented in the figure represents rational decision-making. The more the crossing point moves to the right, the more the human is under-trusting the information provided in the external world and starts to think deeper about the external threats. The amount of cognitive cycles increases and the likelihood for omissions increases as well. The more the crossing point moves left the more a human trusts that the internal world is providing enough capabilities to cope with the external threats. The likelihood of errors of commissions increases. The result for both errors of omissions and commissions is an ogivian-shaped probability function, which depends on the overlap. The curve is valid for omissions and commissions. The uncertainty in both distributions results in the slope of the probability function.

Errors of commission result if the internal world implements results of a thinking process that have no correspondence in the external world. Errors of omissions result if external threats cannot be handled by internal capabilities.

Events do only provide in some particular cases relative frequencies that can be used as estimates for human error probabilities directly, because plant experience is, in principle, not able to provide sophisticated information to calculate a *HEP=n/N*. The reason for this is that every event reporting is always defined by a certain message-threshold that was exceeded (no matter by whom, how, or on which level of detail the event is reported). The message-threshold usually increases with the distance from the working environment. It is relatively low within the business unit, higher if an internal auditor comes and very high, if an external examiner or regulator tries to examine the event. However, the principal problem of event threshold does not disappear even if direct access to the event exists. Hence, the information collected from plant experience may only be taken to support calibration in many cases.

The parameter required for quantification is the amount of overlap between the threats of the external world. Events provide this information: even if the event threshold is high, the rank of criticality of errors persists as valid information. Hence, ranks reflect the behavioural dimension well enough to calibrate data. Figure 6.10B therefore reflects the way to move from events to human error probabilities.

The ogivian-shaped curve can be expressed by several mathematical formulas. The formula of Rasch (1980) was found following the relation based on 220 operational incidents in the nuclear industry validated with 69 items of the THERP method in Sträter (1997/2000). The algorithm could herewith also validate the THERP data for the German nuclear industry (BFS, 2004).

$$P_{\text{Failure of Type}_i} = \frac{e^{\left(\frac{r_i - \mu}{s_n}\right)}}{1 + e^{\left(\frac{r_i - \mu}{n}\right)}}$$

With:

$r_i =$	Percentage of errors in a situation given all situations of the same type in the database.
$\mu = 0.5$	Adjustment of the location of the crossing point (0,5 assigns rational processing).
$s_n \approx 0.075$	Empirical parameter to adjust slope of the ogivian-curve (lower values make the curve steeper).
$e = 2.718$	Natural Exponent, parameter to adjust the ogivian-shape of the curve to the assumed distribution of both worlds.

The approach has the advantage of being logically inferable form the cognitive processing loop. However, other mathematical expressions may be used to represent the ogivian-shaped curve (as long as it provides integration of the normal distributions of the internal and external worlds). Hence, Bayesian methods, binomial approaches or beta distributions can suit the psychological need to integrate internal and external world as well. The essential criterion is that the mathematical approach has to account for the uncertainties in the external world and the internal world as well as the distribution of both. The mathematical approach used here assumes that both distributions are the same. However, as the picture above indicates, the distribution of the external threats might differ so that the model has uncertainties in treating probabilities for errors of commission and errors of omission equally. However, the figure is exaggerating the differences in the shape. Usually the differences between both are small and may not lead to noticeable differences for human error probabilities.

The strength of the approach is that the Laplacian HEP definition comes up as a special case of the ogivian-curve, if the parameters s_n and e are adjusted accordingly. The Laplacian HEP definition is a special instance of the calculus. This suits the law of scientific logic that any new model should contain the preceding approach as a special case.

Quantitative assessment of human reliability is one of the most controversial and difficult parts of HRA. Several decades of HRA-research showed that hunting for HEPs in the classical Laplacian sense seems to be a difficult task that causes data-collection problems that were not solved at all. Consequently, one has to ask: why was the HRA community not able to generate a sufficiently high quality of HEP-collection for over 50 years? Following this observation, is the quantitative assessment of human reliability even in vain?

The answer is: no, but assessing humans is more difficult than assessing technical components, and advanced models have to be used in practice, which need to consider a statistical approach. Such a statistical approach needs to include:

• Context, Situation and Cognition: The quantification method has to consider the internal and external worlds as well as their uncertainties and the unavoidable incompleteness of information of any data source for Human Reliability Assessment.

- A multiple parameter approach, which allows us to represent the dependencies and provides the statistical proof needed to transfer the construct into a quantitative figure (an underlying construct, an operational definition of the construct, an error dimension of the construct, a distribution for the construct, and a one peak, monotone, intransitive shape).
- A statistical approach considering all reference-values for calibration (i.e., a minimal-square approach which allows us to judge the statistical robustness of the figures generated rather than single expert judgement).
- A mathematical approach that allows adjusting the parameter according to the data sources and contains the Laplacian approach as a special instance in the quantification.
- The possibility to replicate the way data were derived (even down to an event) in order to meet the scientific criteria of traceability.

Assessing and Validating the Cognitive Processing Loop

The cognitive control loop described in Part II provides design-recommendations for remote settings based on predictions of human behaviour under given circumstances by using human performance data from operational experience (i.e. events where human behaviour led to incidents or accidents). It is based on experiences and research performed in the nuclear, aviation, automobile design and occupational safety industries (overview in Sträter, 2004b). It demonstrates that technical settings can be evaluated and design recommendations can be drawn proactively. This chapter will also demonstrate how the method allows description and prediction of the cognitive aspects. Examples from incident analysis, as well as from system design in the automobile industry are presented.

The applications show a way forward to overcome the limitations of current safety assessment practice, to better integrate cognitive errors of commission into safety assessment, and to find better mitigation measures for safety critical human behaviour. The following aspects should be demonstrated:

- The assessment of the cognitive resolving mechanisms of expectancy-oriented reasoning.
- The assessment of the cognitive resolving mechanisms of utility-oriented reasoning.
- The importance to consider experiences for the safety assessment of errors of commission and organizational aspects.

The Assessment of Cognitive Expectancy

The discussions in Part II of this book showed how much the mechanisms, which drive errors of commission are stemming from attitudes, expectations, goals and preferences of operators. In order to gain insights concerning the importance of biases (expectations) an investigation in the car industry using the dual-task paradigm was conducted. The subjects had to drive a car and, in parallel, had to perform a simple additional visual decision-response task. Performance measures were taken for the decision task as well as for the driving task. The factor 'expectation' was realized by biasing the possible response frequencies (Sträter et al., 2001).

The secondary task was presented on a display, positioned at the lower half of the centre-console. One of three randomly selected symbols (parallelogram, square and triangle, see Figure 6.11) was successively shown on the display. For normal driving, the symbols shown were outside the fovea, which means that the subjects had to make a fixation to be able to look at the display and to distinguish the shape of each symbol.

Figure 6.11   In-car display with secondary task

The factor 'expectation' was realized by biasing the likelihood of occurrence. If recognizing a parallelogram the drivers had to press the right steering-wheel button (likelihood of ~30% that this response button is correct). The left button had to be pressed if they recognized one of the other symbols (likelihood of ~70% that this response button is correct). After pressing the button, the next symbol occurred. Subjects could decide when to look at the display and how long this deflection from the drive-scenery should last with respect to their accepted individual safety risk level. The point of time when the secondary task was completed and the correctness of this performance were recorded.

Two groups with 11 subjects each (7 female/23 male, mean age 37.4 years, standard deviation 14.1 years, most of them experienced drivers), had to perform the simple visual decision-response task while driving a car under real traffic conditions. Before driving, the subjects had to practice the secondary task in a dry run, which was taken as the reference condition for each driver. The drivers had to drive the same route three times. The first driving trial was used for getting familiar with the car, the route and the experimental situation; the following two trails were the test trials. Participants were told that the driving task has priority and they should perform the secondary task as good as possible in consideration of the driving situation.

The driving distance was 33 km (one-way). A standardized scheme concerning these driving sections was used to classify the traffic scenes (Fastenmeier, 1995). The course included six selected sections with different types of roads (conventional road, highway, city, secondary road) and different traffic densities (rarely other cars ahead vs. often cars ahead) and distraction by traffic signs (low vs. middle distraction). All scenarios were of middle or low complexity. High complexity was avoided in order to avoid potential accidents in this field experiment. In total, 8099 trials with 30 persons were collected and HEP (human error probabilities) were calculated for errors of commission in the decision-response task.

The Human Error Probability (HEP) was calculated for each possible combination of signals and responses (Figure 6.12).

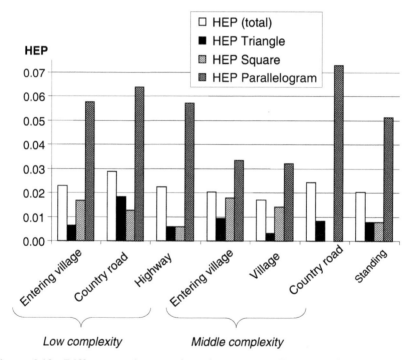

Figure 6.12 **Differences in reaction time responding to different symbols when performing the decision-response while driving and in the stationary condition**

Overall, a HEP of 0.053 was found for the biased condition while a HEP of 0.014 was found for the unbiased condition. No differences were found between the various conditions of complexity of the driving sections (highway, city, conventional) but significant differences were found for the two different conditions that operationalized the bias of persons. There was not even a significant difference between the driving conditions and the standing condition.

In conclusion, the influence of the factor 'expectation' obviously overshadows the error probability one would expect for simple signal response tasks. The investigation showed that the probability for an error of commission based on expectancy is quite high compared with HEPs that would be considered for such tasks if the conditions are treated as independent.

Differences between the varying driving sections due to different driving complexity were significant only in a few cases. There are also just small differences between the different sections and different tasks (parallelogram, square and triangle) observable. Further evaluation also shows that the results regarding the HEP are independent from the available reaction time (see Sträter et al., 2001).

This investigation only gives a first quantitative impression. Many aspects of a task were not considered, like a systematic variation of the number of signals, mapping of signals to buttons, visual type of signals, number of signals and action to be mapped, etc. Further investigations on these aspects would be helpful in understanding how biases are generated in the human mind as well as how to expand the possibilities to assess them in PSA.

The Assessment of Mental Utility

A second setting relates to the assessment of mental utility in early design stages where a product is still in a conceptual state or prototypes are not available. In such cases, the layout of the system may not be known. Only functional requirements are specified so that the information necessary for usability testing and user involvement is missing. A designer's question would be for instance: what is the impact on the driver's cognitive processing and decision-making in critical traffic situations, if I introduce steer-by-wire in the next car generation?

Such a problem was addressed in Theis (2002) for a big automobile manufacturer in Germany. The increasing use of electronic components in the automotive industry requires high reliability, maintainability and safety. The trend of integrating electronic devices started with systems such as the antilock brake systems (ABS) or airbag. Nowadays, electronics affects almost all vehicle components. Increasingly, the electronic vehicle systems are based on electronic control units and electro-mechanical actuators instead of the traditional mechanical and hydraulic systems. In particular, several car manufacturers are developing drive-by-wire systems, similar to fly-by-wire systems in airplanes. In contrast to standard control systems like ABS, such systems may lack mechanical back-ups. To compensate for the lack of a conventional back-up system, other means to assure safety need to be established.

Systems are consequently built as highly redundant as possible and with fallback states, which eventually need to be handled by the driver. To inform the driver about a degraded system state, an alarm strategy needs to be established. One design alternative is a warning to the driver on a system fault combined with a limited functionality of the system until the system is repaired again. The human reaction on system failures is essential for a successful implementation of this alternative. In the case of a fault in the drive-by-wire system, the human is obliged to stop the car immediately for safety reasons. The initial design suggestion was to deal with faulty system states by introducing a strong visual warning combined with an auditive alarm like:

- 'WARNING! Damaged steering, please stop!'

The design question is whether this message leads to compliant behaviour of the driver. Possible human failure modes are:

- precipitate reaction (immediate compliance);
- tardy reaction (late compliance);
- omission of reaction (no compliance).

Prediction of cognitive aspects The CAHR method was used to assess the possible reactions of drivers if a steer-by-wire system fails. The method was used in the way described in Figure 6.9 (above) to classify contextual aspects, probable cognitive coupling and possible reactions of drivers. Cognitive control modes were predicted based on the situational conditions. For the most safety critical combinations, the quantitative path of the CAHR method was used to assess the human error probability (see Table 6.3; according to Theis, 2002). Quantitative assessments were performed to determine the combination of traffic situation (contextual aspect) and control strategy (cognitive tendencies).

Table 6.3   Application of the CAHR method for predicting possible reactions of drivers to failures of the steer-by-wire systems in automobiles

Contextual Aspect	Cognitive Tendencies Cognitive coupling	Fixation	Information ignorance	Expectancy-driven	Information and goal elaboration	Information enrichment
Road quality and traffic-situation	High demanding			0.13		
	Low demanding				0.67	
Situational and environmental factors	High demanding	0.10				
	Low demanding				1.00	
Driving manoeuvre and type of travel	High demanding	0.25				
	Low demanding	0.22			0.22	0.22

The predictions on the combination 'driving manoeuvre and type of travel' related to 'fixation' were considered as being the combination with the most considerable impact on safety and were later on validated in the experiment using two main conditions. Types of travel comprises whether the driver is on a journey where he is forced to reach a certain destination (e.g. the goal to make it to an important appointment) or whether he has an unspecified mission (e.g. weekend excursion). The combination 'ignoring an alarm of the steer-by-wire system due to fixation' was predicted with a probability of HEP~0.25 for high demanding situations and HEP~0.22 for low demanding situations. The prediction is more or less equal for both high demanding and low demanding traffic situations. The quantitative correspondence leads to the conclusion that the human behaviour is less related to the cognitive demand but more to the type of travel (i.e., the goal the driver has).

Driving simulator study The methodological approach and the theoretical results were validated in a driving simulator study with 36 test persons. The test persons had to fill in a standardized questionnaire that included questions concerning their age, gender, their driven mileage etc. Moreover, they had to estimate statements, which were subdivided into six categories in the form of a self-assessment regarding their attitudes towards driving. After the questionnaire was filled out, the test persons were introduced to the vehicle and its features and to the tasks to be performed during driving on a simulator course.

Two conditions were set up in the experimental runs. In the first condition the car had to be operated under the condition of a time-critical highway-trip where the driver had to reach an important appointment for their professional career (i.e., they were induced to show the attitude to focus on a particular aim). In the second condition the drivers were asked to drive as a relaxed holiday-trip (i.e., the drivers were induced not to be fixed on a time-critical aim). The objective of the simulation was to investigate the drivers' reaction in the case of a system failure of the steer-by-wire system (Theis, 2002).

The alarm was set up unambiguously to exclude possible errors due to not perceiving the alarm itself. The alarm system was realized as a 'staggered system'. First, there was only a visual alarm presented providing the instruction to stop the car. Compliance with this alarm was classified as early reaction. Second, an additional haptic alarm was introduced into the steering wheel mechanism. The haptic alarm led to a shaky steering wheel, which gradually increased until the car could hardly be steered. Compliance with this alarm was classified as late reaction. Finally, if test persons did not react on both alarms and continued driving to their destination, this was classified as no reaction. The experiment was performed in a full-scope car-simulator.

The persons reacted as shown in Figure 6.13 (see Theis & Sträter, 2001 for additional discussions). The following failure-modes are distinguished:

- Failure mode 1: driver stops, continues to drive until the tactual feedback is initialized (early reaction on visual alarm).

- Failure mode 2: driver drives on slowly until the end of the experiment (late reaction on visual plus haptic alarm).
- Failure mode 3: driver continues the run until the end of the experiment (no reaction on alarm even with heavy haptic disturbance).

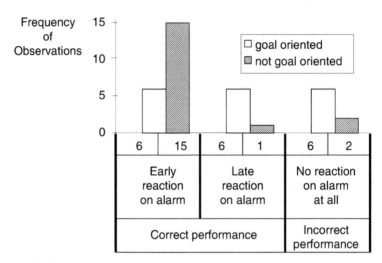

Figure 6.13   Frequency of failure occurrence for ignoring a critical alarm in a steer-by-wire car

In total, 36 drivers participated in the study distributed to 18 drivers in each condition. In the first condition, the drivers had to reach an important appointment (goal-oriented condition). In the second condition, the drivers were asked to make a weekend excursion without any appointment (not-goal-oriented condition).

The diagram shows that six out of the 18 drivers in the goal-oriented condition did not react on the failure at all. In the not-goal-oriented condition, only two out of the 18 drivers showed non-compliant behaviour. Herewith, the experiment revealed the following quantitative HEP-figures for the behaviour:

- for fixation error (goal orientation)   6/18=0.33
- for fixation error (no induced goal orientation) 2/18=0.11
- for fixation error (total)   8/36=0.22

Overall, a probability of HEP~0.22 was observed for fixating on the goal and ignoring information that would contradict or inhibit the accomplishment of the driver's main goal to reach his or her destination. The quantitative result shows that this average figure increases if the goal is more important (33% of the drivers ignored a critical warning and preferred to reach the important appointment). Even in cases where no certain goal had to be accomplished, still 11% of the drivers ignored the critical system failure.

The HEP-figure (total) matches the prediction of the CAHR method for this combination. The same psychological mechanism observed as important in 'driving' a nuclear facility were observed as being responsible for driving a car (which is not surprising for a psychologist since they assume that humans have properties that are independent of the technical system they are involved in).

The results have quite an impact on the design of vehicles, since a safety assessment based on classical models would have revealed a far lower human error probability. The results have a considerable importance for safety assessments in other industries, like the licensing of nuclear facilities or safety cases in Air Traffic Management.

From prediction to design   The experimental setting revealed a corresponding quantitative figure as predicted by the CAHR method for the critical combination (HEP~0.22 was observed; HEP~0.22 to HEP~0.25 was predicted). The study confirmed herewith that cognitive processes observed in nuclear power plant events are transferable to the decision behaviour of a driver in a given situation for both the qualitative approach of cognitive coupling and the quantitative prediction of human behaviour. The same cognitive errors and error mechanism were found in the nuclear events as well as in the car-driving experiment under given situational conditions. Overall, this work shows a successful application of the method outlined in this book, because the qualitative and quantitative expectations fit the experimental ones.

This fact, probably surprising from the engineering perspective, is obvious from the psychological perspective since humans are involved in both settings and, despite all differences between human beings, the cognitive processing loop, the cognitive coupling and the cognitive tendencies appear to be quite similar. The results achieved during the investigation allow using the approach to predict cognitive control strategies in other industrial settings as well, if the description is based on the coupling between human and technical systems.

The CAHR method allows, in a second step, to come up with suggestions on how to improve the design of the technical system (the design path of the approach). The contextual conditions and contributing factors are known for each quantitative figure the method comes up with. The conditions and factors can be ranked according to their importance. Figure 5.3 or Figure 5.6. (above) are examples for such a rank profile for communication and cognitive demands. The ranking for ignoring a critical alarm would lead to the following contributors and design suggestions:

- Time pressure, situational pressure (i.e., to reach a specified goal), and complexity of a situation are circumstances that reinforce this behaviour.
- Improving the ergonomics design of the alarm system cannot eliminate the tendency to ignore the alarm.
- The ignoring could be reduced by improving the explanation on how steer-by-wire systems work, by preparing the driver better for possible scenarios of

failures, and by providing an unambiguous prescription on how to behave in critical situations.

Experiences as an Approach to Assess Errors of Commission and Organizational Aspects

The method and the applications outlined above show the importance to consider the cognitive processing loop, in particular the generic memory architecture distinguishing between experiences and concepts in safety assessments. The approach can also be used for assessing higher-level influences on decisions like organizational aspects. However, as experiences are derived from event information, the succeeding approach to assess errors of commission and organizational aspects implies enhancing event reporting regarding human aspects (e.g. IAEA, 2001b; Eurocontrol-Agenda, 2005). Human Error analysis and prediction means complete coverage of all operational levels.

The relevance of expectations and utility for safety assessments The assessment of cognitive expectancy and mental utility showed how much the cognitive aspects leading to errors of commission are underestimated in safety assessments.

Regarding expectation-related cognitive processing, methods like THERP would assess such simple signal response tasks (one signal/one button condition vs. two signals/one button condition) with a HEP~0.01 for working conditions with distractions (the driving conditions) or HEP~0.001 for static working conditions (standing condition in the experiment). The experiment showed that the assessment of EOCs implies much higher values if biases come into place. The less biased two signals/one button condition shows a probability of about HEP~0.014 in the investigation, which is in the expected range of the rough estimation of HEP~0.01). The biased condition (one signal/one button) shows a probability of about HEP~0.053, which is a factor 5 higher than the rough estimation of HEP~0.01 would have suggested.

The quantitative figure of HEP~0.22 to HEP~0.25 for utility-oriented cognitive processing appears to be quite high compared with Human Error Probabilities used in classical Human Reliability Assessment methods. Nevertheless, the research going on in the field of errors of commission shows that Human Error Probabilities for cognitive behaviour appears to be generally of this order of magnitude (OECD, 2000). This value is supported by investigations in the nuclear industry (see Reer et al., 1999). The classical method THERP (Technique for Human Error Rate Prediction; Swain & Guttmann, 1983) would have assessed this situation with a human error probability of less than HEP~0.001. This value is a mis-assessment by the factor of ~220 (ranging from ~110 to ~330). Such uncertainty has a considerable impact on licensing steer-by-wire cars for public use or any other safety assessment that includes human behaviour.

Both results clearly indicate that probabilistic assessments of cognitive processing have to be performed with an entirely different approach than current assessment methods like THERP suggest. Applying the assessment models designed within the techno-morphologic information processing approach to cognitive errors leads, in the case of expectation, to a mis-assessment of a factor 5 to 10 and in the case of utility-oriented processing even a factor of 200. The factor of 200 was confirmed by studies in Air Traffic Management. A difference of factor 100 was found between safety cases and event-based assessment of human reliability (Cartmale, 2004).

Cognitive errors lead to much higher failure probabilities. Such differences are vital for the credibility of safety assessments, because cognitive errors have the potential to break safety barriers, as Chernobyl or the mid-air collision at Lake Constance showed. However, currently, these mechanisms are not integrated in Probabilistic Safety Assessments (PSAs). The findings can be summarized as follows:

- Cognitive errors have to be assessed in a completely different way than is currently done in safety assessments; one can neither say that the concept of traditional Performance Shaping Factors (PSFs) like procedure or display quality etc are valid factors for a error probability nor should one use the quantitative assessments as a basis of a techno-morphologic approach.
- The results show that qualitative as well as quantitative assessment has to consider the experiences people have gained in the task over their years of working.
- Finally, this observation is a strong reason to abandon the techno-morphologic processing in safety assessments.

Experiences gained with the external world and their result into cognitive errors
As discussed in Part II, violations are a result of resolving mechanisms to overcome constraints in the working environment. They slowly build from first experiences until they are manifested as habits or attitudes towards the system. The context and the duration of experiences of a human need to be taken into account to understand his or her violations. Of special importance are experiences related to compliance to rules and procedures.

Investigation of incidents often conclude 'lack of awareness', 'lack of knowledge' and 'inattention' as human causes for non-compliance. In nuclear energy for instance, events with a human behaviour contribution (or human factors) are attempted to be prevented in 43% of cases by using 'training' as a countermeasure to prevent reoccurrence of such events. Looking at such dominant figures makes one doubtful whether the statement and the concluded measure are correct. A central question of proactive safety is whether such measures are really optimal to prevent errors by the person who could do the error or of other persons under similar conditions.

Figure 6.14 shows the how factors in the external world influence human behaviour. The data-basis for this evaluation was again the 220 events from the nuclear industry, because the interrelations were systematically investigated in this study. It demonstrates for other industries what one can expect to get from the systematic evaluation of events or accidents. The analysis of interrelations shows how much violations depend on the experiences the persons gained with the technical system.

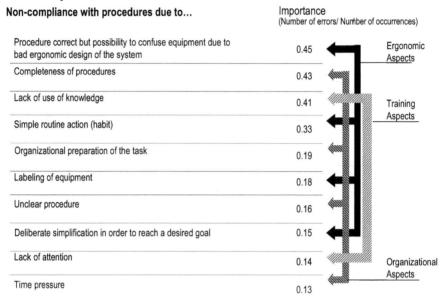

Figure 6.14 **Influencing factors and their interrelations**

The figure shows the different causal factors for the problem of compliance to operational procedures. The classical view relates violations to deliberate simplification in order to reach a desired goal, to lack of attention, or time pressure. As the importance in the figure indicates, these three, however, actually explain the least of the observed events. One can find the standard-factor 'lack of attention', but on the least ranks rather than on the first ranks.

One can find on the first ranks of the causal factors: possibility to confuse equipment due to bad ergonomic design, completeness of procedures, and lack of knowledge about the procedure.

The consideration of several events reveals a systematic pattern of factors leading to the behaviour 'Violation of procedures'. Violations appear to be related, to a great extent, to the context of the person who does the error. In view of several events, it also appears that it is much more efficient to consider a change of the ergonomics of the technical system instead of changing procedures or considering training or instructions. The reflection of operational experience consequently reveals that the countermeasures, like 'improving procedures' or 'training', are expected to be less effective.

Note that this evaluation of events is related to nuclear experiences. The weights as well as some factors shown in the figure may change in an evaluation of accidents in other industries. However, the approach for improving the knowledge about better prevention measures can be transferred. A look into other technical disciplines shows that counter-measures are also very often not properly related to the causal factors of the events.

Assessment of organizational aspects with impacts on decision-making   The potential of evaluating the constraints in cognitive processing numerically by using event-data enables the assessment of the impacts on decision-making. Potential impacts are critical accident situations (accident management) or organizational aspects (including safety culture issues).

Accident management situations are different from normal operation. They describe highly degraded system states, like in fire events or loss of essential safety features of the system. Accident management tasks comprise complex operator actions under considerably unusual system conditions. These conditions may involve, in the nuclear industry for instance, entering different locations in the plant, involving security and fire brigades, coordinating between different locations, coordination with outside authorities, appearance of emergency response teams etc.

Many conflicts of operational goals arise from such situations, which are usually not considered in safety assessments. In the Davis-Besse Event, it was observed that operators changed their goal from the initiation of a required safety procedure (called feed and bleed, which makes the plant unusable in the future) to a more operational goal to keep the plant usable after the event (Reer et al., 1999).

The underlying assumption for reducing the error-spectrum is that operators are normally well trained and differences in the operators' recognition of the cognitive level are compensated by training and qualification. This assumption cannot hold for Accident Management situations. The main reasons are the unknown level of stress and potentially hazardous conditions for performing local actions. Assessment methods based on the techno-morphologic information-processing model (like THERP) are not suitable to deal with the special demands under such plant conditions. The assessment of cognitive aspects in such situations needs careful consideration of decision critical parameters, like time budgets and the experiences operators use in decision-making.

Incidents and accidents of complex systems also show the interrelationship between staff and organization. They present a window to see parts of safety management in 'action' under the circumstances of an accident and can create a new understanding about 'safety culture'. Accordingly, events can be used in a similar manner to generate predictions of organizational aspects for safety assessments. First experiences with the event-oriented safety culture assessment show this as a promising approach (Balfanz et al., 2004). Based on these experiences, any constraint leading to decision problems of the user at the sharp end may be included into safety assessments.

Chapter 7

Integration of Cognitive Performance

The same procedure used to include organizational aspects into safety assessment could be used for aspects such as safety culture and the extra-organizational influences on the culture (e.g. via regulatory issues). So far, only self-assessments of the utilities are considered as an effective approach to address safety culture (e.g. IAEA-INSAG 13). However, self-assessments need a reference (e.g. to decide whether important safety aspects are missing). By including such aspects into the scope of a PSA enhances the safety management and makes risk monitoring more effective.

This chapter suggests an integrated treatment of cognitive aspects in safety assessment from design to operation by elaborating:

- The classical treatment of cognitive aspects in safety assessments.
- An integrated approach of assessing cognitive aspects in design and operation.

The Classical Treatment of Cognitive Aspects in Safety Assessments

Limitations of the Current Representation of Cognitive Aspects in Safety Assessments

Usually, safety assessments are performed according to a certain initiating event and its potential consequences. First, a functional model of the system to assess is created. A certain system failure, which might have potential hazardous effects, is then identified from the system design (e.g. a leak in a pipe in a nuclear power plant or a loss of an engine of an aircraft). The system behaviour after the initiating event is modelled and design measures are taken to reduce the risk of hazardous results, either by increasing the reliability of the system or by establishing barriers to prevent further development of the hazardous system state.

Sometimes this model is rather qualitative (e.g., a list of potential failures and consequences in a table, like in the HAZOP approach). In a more elaborated form, the safety assessment model contains an event-tree to model the sequence of the development of an unwanted event (time-line), a fault-tree to model the possible faults in the system, and a human performance model to model human reliability in a certain failure condition. Overall, this framework can calculate the general safety performance of a system in terms of accidents per time, like severe accidents per year in the nuclear industry or collisions per flight hour in air traffic management. Figure 7.1 presents this general process.

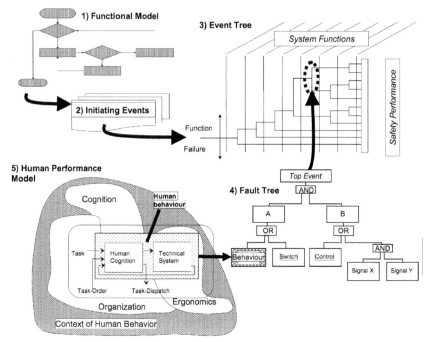

Figure 7.1   The safety assessment model from initiating event to safety performance

The figure makes obvious that current safety assessment models treat humans as components in the system that have to perform a certain task according to the safety assessment model. Many authors have described the inadequacy of the framework for including cognitive aspects (Hollnagel, 1998; Cacciabue, 1998; Siu, 1994; Dang, 1996). However, it is still in use because of a lack of alternatives and historically founded sluggishness towards change. Lacking are, for instance, time dependencies of the actions or the dependencies due to the dynamics of several actions to be performed by the same person. The most critical point regarding classical safety assessment models is that they are not sufficiently prepared to describe or predict the active human roles, which are required in the so-called beyond design-based scenarios of accident management for instance. In these situations, the human is driving the event sequences rather than being driven by them. Such interventions depend on the goals a person or group of persons want to reach. These observations lead to the following conclusions (GRS, 1999; Müller-Ecker et al., 1998):

*   The event- / fault tree technique allows us only to describe a static snapshot instead of the dynamics of a system.
*   The dependencies between persons or within person cannot be modelled satisfactorily.

- The integration of active human involvement (errors of commission) is not possible with these approaches.

Figure 7.2 expresses this problem. Each error of commission would require another event-tree/fault-tree model to be built, which probably also includes another error of commission and so on. Consequently, the current safety assessment methods would require an endless analysis effort to assess the safety performance if active human behaviour should be included.

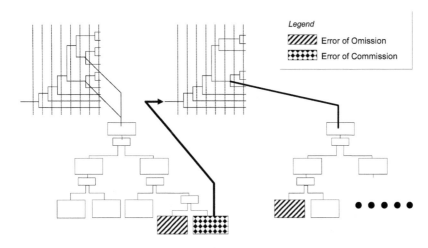

Figure 7.2 Problems of the event sequence approach for including active human involvement (e.g. errors of commission)

These problems of including the cognitive aspects of human errors lead to the conclusion that human errors cannot be treated like equipment failures in safety assessments. The preceding section also showed that the quantitative representation may misjudge the human-induced risk due to such a wrong treatment of human behaviour by a factor of 100 to 200. People have attitudes, constrains and particular properties in information processing, which are not reflected in the classical risk assessment models.

The starting points for better assessment are a couple of more cognitively oriented models recognizing the dynamic nature of human interventions, like:

- CM: Confusion Matrix approach (Potash et al., 1981).
- CES: Cognitive Environmental Simulation (Woods et al., 1987).
- ESAT: Expertsystem zur Aufgaben-Taxonomie (Expert-System for Task Taxonomy; Brauser, 1992).
- MIDAS: Man Machine Integration Design and Analysis System (Corker & Smith, 1993).

- OPPS/SAINT: System Analysis Integrated Networks of Tasks (Wortman et al., 1978).
- COSIMO: Cognitive Simulation Model (Cacciabue, 1992).
- IDAC: Information, Decision, Action in Crew context (Chang & Mosleh, 1999).

They all address cognitive decision-making in dynamic situations and how to link them into safety assessments (Siu, 1994) and are therefore grouped as dynamic risk models (further discussions are in Trucco & Leva, 2004). The Confusion Matrix approach is consistent with the classical safety assessment model and hooks cognitive decisions on this conventional approach. The other approaches suggest a specific approach to model human behaviour. Some disadvantages of the methods are that they did not thoroughly overcome the limitations of the techno-morphologic information processing approach. In some of them short-term memory is separated from long-term memory or the ladder-model is used to model cognitive processes, which neglects the influence of higher-level goals on skill-based behaviour. Due to the reliance on the techno-morphologic modelling of cognitive activities (information processing stages), the methods did not achieve a stable ground for generating suitable data and still use classical quantification approaches like time reliability or expert judgement methods. Herewith the methods ran into the data-problem of not having enough validated data to be used in safety assessments or of being not distinct enough from classical approaches to give a reason to change the paradigm towards dynamic risk modelling.

An Integrated Approach of Assessing Cognitive Aspects in Design and Operation

Dynamic risk modelling needs a model that suits the computational effectiveness of human cognition in real situations. The cognitive processing loop as presented in the preceding chapters suggested an approach to overcome the limitations in the techno-morphologic processing loop in order to overcome limitations in safety assessments. This suggestion contains an approach for data-generation that is logically and empirically followed from the cognitive processing loop, a full integration of techno-morphologic information processing models as special instances of the cognitive processing loop, and an integration of additional cognitive issues not yet considered in safety assessments. Examples are the interrelation of long-term memory, memory span, decision-making and reaction time under dynamic task conditions or the link of dissonance biases and emotional aspects. The cognitive processing loop also revealed a classification of cognitive error-types related to cognitive dissonance.

The cognitive processing loop can hence be understood as a proposal for dynamic cognitive risk modelling that can be applied to a functional model of a system. Explicit and implicit modelling may be distinguished. A real time calculus for cognitive performance is possible using the concept of dynamic binding (computational effectiveness). Representation of the effect of context on cognition is achieved by the coupling of the internal and external worlds. In terms of safety assessment, the external world relates to externally observable information while the internal world relates to the cognitive abilities of humans. The cognitive abilities are represented in the cognitive processing loop either explicitly or implicitly. The explicit aspects are those having an impact on the functions of a system. These are the human experiences (i.e., the memory architecture with its impact on decisions and biases) and resolving mechanisms. The explicit aspects lead directly to the availability of a system function and are hence connected to a certain failure state of the system or to the commission of an inappropriate system function. The implicit aspects are part of the processing loop and show up in the decisions and reaction times. The implicit modelling aspects of cognitive performance (like emotional aspects, memory span, reaction time) do not have a direct impact on the functions of a system but trigger the cognitive processes.

How the dynamic model can be used to model the drivers' behaviour during changing a lane on a highway is shown in Figure 7.3 (according to Schemmerer, 1993).

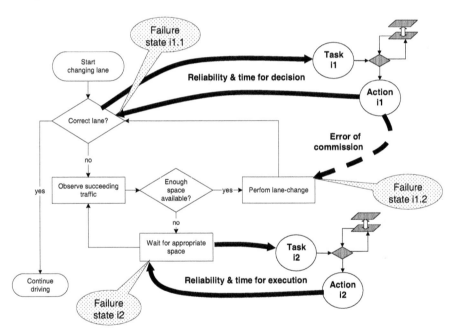

Figure 7.3 **The use of the cognitive processing loop in dynamic safety assessments**

In a functional model view, the cognitive processing loop provides the way a human being sets-up the links between the functions of the system based on his experiences. The dynamic involvement of the human plus his or her own creativity can be modelled by this approach. Compared with the classical inclusion of human behaviour in safety, the efficiency of the resolving mechanisms and the impact of time constraints on the memory are the main additional aspects that can be assessed as contributors to the overall system risk. Errors of commission are represented as deviations from the supposed functional role of the human (Hollnagel, 2004). The deviance from the optimal fit to the functional model can be expressed by the Hamming distance (i.e., the number of deviations from the optimal functional path; Ameling, 1990).

The modelling approach is generic and can be applied to virtually any system, as long as a functional model of the system can be built. A computerized tool called COSI (Cognitive Simulation) was established to show the potential of the dynamic modelling for design and risk assessment (CAHR, 2005). The tool uses UML (Unified Modelling Language) to build the functional model and uses the CAHR approach to model the human behaviour in the functional model. Several instances demonstrate how the system can be used to assess the impact of human behaviour on system safety under dynamic situational conditions. Examples were provided on mental demands and memory load during car driving while operating an additional system, the accident management behaviour of an operator in a nuclear power plant during emergency scenarios, or the communication between a pilot and an air traffic controller. As human behaviour in virtually any system can be represented, the approach would also allow us to model organizational aspects of higher-level decisions in management or regulation.

Subotic & Oeching (2005) are currently investigating the potential of human recovery during system-failures, based on dynamic modelling, and are validating the prediction by a simulator study in Air Traffic Management.

Such a dynamic risk modelling allows actively using risk assessments in the safety management of a system. Rather than providing a static risk picture once at the beginning of the life-cycle of a system, risk monitoring leads to a risk-informed or risk-based decision on changes in the system configuration while the system is running (on-line). A risk monitor tool allows validating that changes of the system configuration (technical as well as organizational) are not harming the overall safety balance and essential safety nets. Shepherd (2002) investigated the potential use of risk monitoring in the nuclear industry and found that the primary benefit of risk monitoring is the cost-effectiveness in aligning safety requirements with operational safety requirements, in particular regarding:

- Prevention of the unsafe on-line maintenance: Risk monitoring allows one to judge whether it is safe to carry out maintenance when the system is on-line. Such a feature would have prevented the mid-air collision at Lake Constance where the system was operated with two safety nets out of service. The risk impact of this change was not aware to the controllers, the technicians or the management. A risk-monitor would have provided the means of controlling the system configurations and ensuring that the overall risk is within the limits.
- Prevention of unsafe planning of system changes: System changes are not ad-hoc actions. They need to be planned sometimes with long lead times. Planning assumes a certain system configuration of those parts of the system that should remain unchanged. However, this might not always be the case. In Air Traffic Management, the strategic airspace flow and capacity management

are planned one year ahead. The planning hence cannot take into account small changes in the operating room equipment that might be introduced during the year. In the nuclear industry, a fuel exchange may need two years of planning. The long lead times postpone safety considerations and therefore may induce constraints and safety issues for the humans at the sharp-end (Hollnagel, 2004). Bad planning runs the risk of 'fining' the sharp end that much that the humans cannot handle the task any more and commit errors. A risk-monitor would allow assessment of the risk in such long-term planning activities and plan system changes safely.

- Prevention of ill-informed accident management: Systems may spontaneously be disturbed leading to undesired risky situations. The safety relevance of such system failures is often not immediately recognizable by operational staff or management so that a trade-off is made between safety and economy towards economic considerations. Such trade-offs may lead to decision errors and errors of commission, as the mid-air collision at Lake Constance showed as well as other accidents in other industries (e.g. Chernobyl). A risk-monitor would allow judgement about the effect of system failures as well as unplanned maintenance or repair requirements on safety and allow drawing the safest strategy to overcome the unsafe state.
- Prevention of neglecting safety in economic planning: Overall, a risk-monitor would provide a basis for planning cost effective system safety by providing risk information in a form that is readily understandable and can be used to demonstrate the level of safety of the system. Risk monitoring would be even more powerful if it were linked with other planning tools, such as planning tools for long term investment (see Eurocontrol-Planning, 2003 for an investment planning tool in Air Traffic Management).

The Importance of a Proactive Design and Assessment

This book offers an integrated view on cognition and safety. It provides a generic cognitive processing loop describing human reasoning and errors. The generic memory structure, distinguishing between concepts and experiences, was elaborated as an important aspect to understand biases as well as goal and utility-oriented behaviour. It provided a coherent picture of human cognition in safety from mental workload via reaction times and induced workload to decision-based human errors in accidents. The issue of errors of commission could be approached from the qualitative analysis point of view as well as from the inclusion of errors of commission into safety assessments.

The generic method described in this book may improve the current way cognition is treated in safety assessment and early design. It presents an approach of design and assessment that starts with human cognition and not with the technical system. Examples of nuclear, automobile, aviation and air traffic management were used to illustrate this path. It was also shown that the approach addresses human characteristics on all organizational layers involved in the design and operation of a technical system, like the designer, operational staff, management, and regulators. Two important possibilities are offered here: a cross-industry view on cognitive science and an approach to overcome hindsight on human cognition.

A Cross Industry View on Cognitive Science

Often, cognition science is seen as an industry specific add-on of the technological development. This view is historically grown from an industry-specific development of technical components. This book contrasted the possibility of an across-industry view on the topic by describing a generic approach of human characteristics involved in the safety design and operation of a technical system in a wide range of industrial areas. In all technical areas (like occupational safety, transport, aviation, process industry and medicine), one can observe a number of similarities and differences regarding human error and their circumstances (Sträter & Kirwan, 2002).

It was shown, for instance, that predictions of cognitive performance of operators in nuclear power plants can be used successfully to predict cognitive performance during car driving. It can be concluded that transfers of cognitive control strategies are possible with this approach, independently from the technical system. This book could only describe Air Traffic Management, nuclear industry, ground-transportation (car, rail), and occupational safety, but similar mechanisms and errors can be found in any other working environment, such as medicine for instance (e.g. Helmreich & Merritt, 1998).

Incidents often have many parallels across industrial settings. A human error while re-establishing a diesel generator in a power plant may have identical mechanisms to an error of a pilot during landing or an air traffic controller in controlling aircraft. Up to now there has been no systematic exchange or learning across the different industries – although serious events in each industry show that such a knowledge exchange is an important issue to investigate. Serious human errors due to loss of corporate knowledge have been observed (IAEA, 2001a). Also in Germany, due to the political decision to end nuclear power generation, considerable problems have occurred due to loss of knowledge. An interdisciplinary knowledge management system could have prevented some events that occurred and that are going to occur.

It could be shown that human errors and their treatment in safety assessment or incident analysis are not dependent on the industry in which a human is working. On the other hand, it was shown that analysis or assessment looking at the component level rather than human characteristics has to fail. The cognitive processing loop provides a framework to exchange commonalties between industries. A generic model of cognition also serves as a knowledge management tool for cross-boundary learning on cognitive constraints and decisions.

A trans-industry comparison of commonalities and differences in human error mechanisms is a powerful knowledge-source to learn from the incidents and experiences of others and to avoid human errors across various industrial settings. It is essential to learn which experiences human have, how they use them in decision-making and how they react under the constraints of the system.

After a serious accident, it is often stated that this is a single human failure, almost beyond control of any design precaution. This is an excuse rather than realistic. The model outlined in this book demonstrated how similar severe accidents in different industries are from the cognitive perspective. A common pool of incident data would allow better analysing of such 'coincident' events and reveal their systematic pattern regarding human cognition. Such an effort would allow one to handle dependencies better between errors and causes and to generate better frequencies and probabilities for human errors, which is needed for better design.

The essential advantage lies in the long-term use of the information as a basis for decisions on investments for precaution measures. In the long term, the additional effort is easily compensated because a systematic analysis and a statistical evaluation of events are worth it in order to learn from other events and to plan better and realize precaution-measures. It enables us to:

- Understand better the behaviour of humans in their own working environment: Each event represents the experiences the humans involved in the accident have collected over the years.
- Find commonalties between different events and herewith be able to distinguish between systematic and random errors: An error observed may not have resulted in a severe accident in one industry but may have resulted in an accident in another industry and hence should be treated seriously.
- Check for completeness of mitigation measures: A mitigation measure introduced may be incomplete or even wrong because of unconsidered side effects.
- Judge about the effectiveness of precaution-measures: A precaution-measure expected to be successful is going to be introduced in one system but has already failed in other industries.
- Have a profound basis for investment-decisions: Investments into a particular measure (like training) can be assessed regarding effectiveness by looking at the efforts others have experienced.
- Define future developments: Common research topics can be identified.

Overcome Hindsight on Human Cognition

Considering human cognitive aspects is often seen as a post hoc approach. This means that, the cognitive issues are investigated relatively late in the design process of a system and then in a 'hindsight' manner. A common approach for proactive assessment and retrospective analysis of human cognition is able to overcome this position and to develop a proactive involvement of the Human Factor.

Task analysis, as the main methodological approach within the Human Factors community, is already a capitulation of Human Factors towards technical design. It acknowledges that functional requirements are the ones to be given in order to be able to include Human Factors aspects appropriately. This is already the first mistake of treating human cognition properly. The design should start and cooperate in virtually any technical design stage in order to be recognized successfully.

The hindsight-strategy is, in principle, an unlucky one for the Human Factor domain. A more or less expensive experiment or usability test is performed to test out the ergonomic quality of a product and the Human Factor persons only have the opportunity to give a go (well done design) or a no go (poor design). A no go in such a usability test then usually leads to the bad image of Human Factors because it judges negatively the work done by designers or other non Human Factor persons. Such a judgement delays the finalization of a product and leads to cost intensive re-design processes. Therefore, Human Factors is more a source for potential conflicts in both the process of creating product and the personnel relationships of designers and Human Factors specialists rather than improving the system.

A proactive approach is, in particular, needed in the field of PSA/HRA (Probabilistic Safety Assessments/Human Reliability Assessment). In PSA, system engineers usually build the safety model with little consideration of the human cognitive aspects. Once the safety model is complete and only the human part is missing, HRA experts are asked to look on the existing model and to make their assessments. However, HRA experts may request changes to the entire safety model itself. Typically, this is the case if considerable errors of commission have been included. Some human reliability issues may even change the basic assumptions of the PSA model (Sträter, 2001b). Recognizing this mismatch too late in the PSA/HRA process means loss of money, less efficient use of the work-force and finally, bad or even wrong safety assessments.

One reason for such a mis-aligned cooperation of technical design and cognitive design may be that different parties are involved in the system design (or assessment) and that these parties have a different education as well as different attitudes, point of views, group thinking and other factors making a cooperative working style difficult. Another reason may be that the education of a psychologist is usually not well adjusted and aligned to the needs of system engineering. Human Factors is still not a self-standing educational degree in many countries and interdisciplinary work is still extremely difficult because of such 'language problems'.

The outcome of such conflicting situations is that Human Factor aspects are usually under-represented in the assessment or design. This is an additional reason for the under-representation in the error of commission problem in PSA. For decades, PSAs have been constantly failing to address this issue because it does not fit to the classical way of PSA thinking (OECD, 2001; OECD, 2002). The effect is that the risk of nuclear installation has been mis-assessed for many years. It will hopefully never be tested by a serious event somewhere in such installations whether the nuclear industry can afford this mis-assessment. Other industries like, Air Traffic Management, currently build up their safety assessments. They hopefully take a better path and so overcome the development of safety assessments into paper exercises for regulatory purposes that have little representation of the real constraints with which the staff are working.

Chapter 8

Perspectives

Understanding cognition will be of considerable importance for the design of the future working environment. It is to be expected that the current development into distributed workplaces, highly computerized systems as well as highly connected and interdependent exchange of information, will increase further. In particular, the operational phases and operational levels will be spread throughout the global economic system. Liability and responsibility will change due to this development while the system is getting more sluggish and economically driven. The competence and understanding of people working for the safety of a system will not have the same cultural background any more. The nuclear accident in Paks was mentioned as an example where the regulator and manufacturer from abroad played an important role in accident development. The regulator was operating under reduced personnel resources. He was also unaware about the constraints of the manufacturer. The constraints of both regulator and manufacturer were not taken into account by the plant staff or by safety assessments because they appeared to be too remote from the real operational threats.

Economic constraints of such highly distributed and interdependent systems combined with competence and experiences of people on all operational levels will increasingly affect system safety. Only the consideration of cognitive aspects in safety assessments will allow future accidents to be overcome. Current safety assessment is not prepared to deal with such types of distributed working environments and the cognitive aspects resulting from them.

Complex systems are, by definition, complex. Therefore, a book about cognition, which elaborates on one of the most complex systems we know of, is necessarily also complex. This book provides an integrated model of cognition without reinventing the wheel. It integrates established knowledge about this issue and several findings on cognitive processing.

A cognitive processing loop is introduced in this book, which provides an approach to understanding the principal cognitive aspects that couple a human's internal world with the external world. The cognitive model encounters the following essentials:

- Although human cognitive performance is a highly parallel process, a cognitive act appears to be uni-dimensional, i.e. a stable relationship in the cognitive coupling of the internal and external worlds. This desire for consistency is limiting cognitive human performance.

- The internal cognitive world of humans (e.g. goals and attitudes of operators) has a decisive impact on human behaviour. Goals influence what to see and what to do. Habits, heurisms and skills influence what to see and search for.
- Experiences of humans are decisive in understanding the goals and attitudes of humans. They lead to mental utility in working environments.
- The cognitive processing loop is equivalent for human-machine interaction as well as group interaction.

The approach achieves an increasing the spectrum of human cognitive characteristics that are often not considered for the safe design of systems. Regarding the treatment of cognitive performance, such an integrated view on cognition allows us to

- Support design and operation of systems.
- Model human contributions to safety on all operational levels.
- Model Human Reliability Assessment and human induced failure states (errors of commission).
- Use retrospective analysis of human behaviour in prospective safety assessment.
- Link of human factor aspects (like cognitive workload, situational awareness) with safety assessments.

The focus of the book is on showing how an integrated view can be used to consider cognitive human contributions to risk in the design of better automation. Automated systems and the human part of any system are dependent entities. Only the integration of human characteristics with automation design enables system safety and efficiency.

Sheridan (2002) mentions a number of impacts of automation on human cognition and human society. Amongst them are threatened unemployment, erratic mental workload, centralization combined with de-localization, skill degradation, technological illiteracy and trust. In addition, the aspect of accountability is mentioned as intimidation (greater responsibility because the systems operated by one person are getting bigger). Abandoned responsibility and enslavement are mentioned as important dangers to be aware of.

Unresolved concerns exist in situations where a human has to decide whether he leaves a certain task to an automated system or whether he takes it over himself. Who will be responsible if an automated system fails and the human fails to take over the responsibility in time? In particular, the aviation and nuclear industries suffered from such questions. Therefore, ambiguity is known as leading to so-called errors of commission of humans with serious incidents. Humans are forced in cases of unclear system behaviour or unclear responsibilities to make judgements under uncertainty and cope with the uncertainty by using heurisms and eventually rule-breaking behaviour if the system procedures are badly designed.

Successful accident prevention therefore requires a thorough evaluation of human cognition from the operational to the management level and from design to operation. There is the need to improve understanding of the cognitive processes of humans in critical situations and to provide a proposal to overcome a guilt-related view on cognition, which has also impacts on legal considerations about human error. Besides the legal aspects, such a guilt-free and blame-free understanding of cognitive characteristics is essential because the same human characteristics of decision-making and intelligence are decisive in successfully dealing with those accidents a technical system cannot cope with any more. Considering that it takes about 10 years until research results into an effect in applied industrial use there is not much time left to decide on well-advanced cognitive paradigms for design and assessment in order to meet the safety threats of future technological developments.

Finally, the integrated cognitive modelling suggested in this book enhances the common understanding between all involved parties in system design and operation and eventually may lead to better recognition of cognitive issues in human   society.

Bibliography

Adams, J. A. (1982). Issues in human reliability. *Human Factors,* 24, pp. 1-10.

Airsafe (2002). *Fatal Events Since 1970 for British Midland.* AirSafe.com, LLC <http://airsafe.com/events/airlines/midland.htm> (01.02.2005).

Airsafe (2004). *27 March 1977; KLM 747-200; Tenerife, Canary Islands.* AirSafe.com. LLC <http://www.airsafe.com/events/models/boeing.htm> (01.02.2005).

Amalberti, R. (2001). The paradoxes of almost totally safe transportation systems. *Safety Science,* 37, pp. 109-126.

Ameling, W. (1990). *Digitalrechner-Grundlagen und Anwendungen.* Braunschweig: Vieweg.

Anderson, J. R. (1983). *The Architecture of Cognition.* Cambridge: Harvard University Press.

Anderson, J. R. (1996). *Kognitive Psychologie.* Heidelberg: Spektrum Akademischer Verlag.

Attneave, F. (1974). *Informationstheorie in der Psychologie: Grundbegriffe, Techniken, Ergebnisse.* Stuttgart: Hans Huber.

Baddeley, A. (1982). *Your Memory: A Users Guide.* London: Sidgwick & Jackson.

Bainbridge, L. (1987). The Ironies of Automation. In: Rasmussen, J., Duncan, K. & Leplat, J. (Eds), *New Technology and Human Error.* London: Wiley.

Balfanz, H. P., Linsenmaier, B. & Straeter, O. (2002). Development of Practical Criteria for Safety Culture Assessment in German Nuclear Power Plants. *PSA 02,* American Nuclear Society, 6-10 October 2002, Detroit.

Balfanz, H., Junge, R., Linsenmaier, B., Rausch, H., Sträter, O., Kallenbach-Herbert, B., Sauerbrey, U., Sickert, M. Oehmgen, T. & Rotzsche, J. (2004). *Entwicklung praxisgerechter Bewertungskriterien für die Sicherheitskultur in deutschen Kernkraftwerken.* Salzgitter: BFS.

Baret, H. & Vermeiren, K. (2004). Experience in developing and implementing the controller's working position in ATC operations. *Human Computer Interaction,* HCI Aero, 2004, Toulouse.

BAUA (2002). *Sicherheit und Gesundheit bei der Arbeit 2002.* Dortmund: BAUA (see also www.baua.de).

BFS (2004). *Methoden zur probabilistischen Sicherheitsanalyse für Kernkraftwerke.* Facharbeitskreis Probabilistische Sicherheitsanalyse für Kernkraftwerke. Salzgitter: BFS.

BFU (2004). *Investigation Report AX001-1-2/02-May 2004.* Braunschweig: Bundesstelle für Flugunfalluntersuchung (BFU) [see also www.bfu-web.de].

Bochenski, I. M., (1954). *Die zeitgenössischen Denkmethoden.* 3. Auflage 1965, Bern (CH): Francke Verlag.

Bonini, (2004). *A Model of Trust in the Work of an Air Traffic Controller.* [Dissertation, Department of Psychology, University of Dublin].

Brauser, K. (1992). ESAT-Ein neues Verfahren zur Abschätzung der menschlichen Zuverlässigkeit. In: Gärtner, K. (Ed), *Menschliche Zuverlässigkeit*. DGLR-Bericht 92-04. Bonn: DGLR.

Broadbent, D. (1958). *Perception and Communications*. New York: Pergamon Press.

Bubb, H. (1992). *Menschliche Zuverlässigkeit*. Landsberg: EcoMed.

Bubb, H. (1993). Systemergonomie. In: Schmidtke, H. (Ed). *Ergonomie*. München: Hanser.

Bubb, H. (1999). Möglichkeiten und Grenzen des Mensch-Maschine Systems. *4. Expertengespräch Mensch-Maschine-Wechselwirkung*. BFS KT-22/99, 14.4.-15.4.1999. Salzgitter: BFS.

Bubb, H. (2001). Menschliche Zuverlässigkeit und Arbeitsschutz: Eine Möglichkeit für effizienten Unfallschutz. *27. Internationaler Kongress für Arbeitsschutz und Arbeitsmedizin Innovation und Prävention 2001 (A+A)*. Düsseldorf: BAUA.

Bubb, H. (2002). Computer aided tools of ergonomics and system design. *SAE'02*, Detroit: SAE.

Bubb, H. & Sträter. O. (2005). Grundlagen der Gestaltung des Mensch-Maschine-Systems. In: Zimolong, B. & Konrad, U. (Eds), *Enzyklopädie der Psychologie: Ingenieurspsychologie*. Göttingen: Hogrefe.

Bubb, H., Sträter, O. & Linsenmaier, B. (2004). Vorüberlegungen zu der VDI-Richtlinie 4006. Blatt 3: Menschliche Zuverlässigkeit: Ereignisanalyse. *46. Fachausschusssitzung DGLR T 5.4 Anthropotechnik*, DGLR, 12.-13.10.2004, Warnemünde.

Bunting, A. (2003). *Integrated Performance Modelling Environment*. QinetiQ, Centre for Human Sciences, Aldershot (UK).

BWB (1996). *Handbuch der Ergonomie*. München: Hanser.

Cacciabue, P. (1992). Cognitive Modelling: A fundamental issue for human reliability assessment methodology. *Reliability Engineering and System Safety*, 38, p. 91.

Cacciabue, P. (1998). *Modelling and Simulation of Human Behaviour in System Control*. Heidelberg: Springer.

CAHR (2005). *Communication-Assessment-Human-Research* (see also www.cahr.de).

Card, S. K., Moran, T. P. & Newell, A. (1983). *The Psychology of Human-Computer Interaction*. Hillsdale (NJ): LEA.

Carnegie, D. (1999). *Freu dich des Lebens*. Berlin: Ullstein.

Cartmale, K. (2004). Lessons learnt from incidents investigated in Maastricht Upper Area Control Centre. *Safety Team Meeting 21*, Eurocontrol, 2004, Buchareast (Romania).

Chang, Y. H. & Mosleh, A. (1999). *Cognitive Modelling and Dynamic Probabilistic Simulation of Operating Crew Response to Complex System Accidents*. College Park (US): University of Maryland, Center for Technology Risk Studies.

Chapman, L. & Chapman, J. (1969). Illusory correlation as an obstacle to the use of valid psychodiagnostic signs. *Journal of Abnormal Psychology*, 74, pp. 193-204.

Chomsky, N. (1965). *Aspects of the Theory of Syntax*. Cambridge (MA): MIT Press.

Comer, M. K., Seaver, D. A., Stillwell, W. G. & Gaddy, C. D. (1984). *Generating Human Reliability Estimates Using Expert Judgement*. NUREG/CR-3688, 1. Washington DC: NRC.

Comer, R. (2001). *Klinische Psychologie*. Heidelberg: Spektrum Akademischer Verlag.

Coombs, C. H. (1964). *A Theory of Data*. New York: Wiley.

Corker, K. M., & Smith, B. (1993). An architecture and modelling for cognitive engineering simulation analysis: application to advanced aviation analysis. *9th AAIA Conference on Computing in Aerospace*, AAIA, 1993, San Diego.

Dang, V. (1996). *Modeling Operator Cognition for Accident Sequence Analysis: Development of an Operator-plant Simulation*. MIT, Cambridge (MA) [PhD Thesis].

De Groot, A. D. (1966). Reception and memory versus thought: Some old ideas and recent findings. In: Kleinmutz, B. (Ed), *Problem Solving: Research, Method and Theory*. New York: Wiley.

DeKleer, J. & Brown, J. (1983). Assumptions and ambiguities in mechanistic mental models. In: Gentner, D. & Stevens, A. (Eds), *Mental Models*. Hillsdale (NJ): Erlbaum.

Doppler, K. & Lauterburg, C. (2002). *Change-Management: Den Unternehmenswandel gestalten*. Frankfurt: Campus Verlag.

Dörner, D. (1976). *Problemlösen als Informationsverarbeitung*. Stuttgart: Kohlhammer.

Dörner, D. (1997). *Die Logik des Misslingens: Strategisches Denken in komplexen Situationen*. Hamburg: Rowohlt.

Dörner, D. (1999). *Bauplan für eine Seele*. Hamburg: Rowohlt.

Dörner, D. & Schaub, H. (1994). Errors in planning and decision-making and the nature of human information processing. *Applied Psychology: An International Review*, 43 (4), pp. 433-453.

Dorsch, F., Häcker, H. & Stapf, K. H. (Eds) (1994). *Dorsch Psychologisches Wörterbuch*. Bern: Hans Huber.

Dougherty, E. (1992). Context and human reliability analysis. *Reliability Engineering and System Safety*, 41, p. 25.

Drozdowski, S. (2003). *ACAS RA Downlink Feasibility Study*. Brussels (Belgium): Eurocontrol.

Eberlch, L., Neugebauer, C. & Sträter, O. (1989). *Modelle des Wissens und Methoden der Wissensakquisition: Eine Übersicht*. Arbeitsbericht Nr. 1-55. Aachen: Institut für Psychologie der RWTH Aachen.

Edwards, W. (1977). How to use multi-attribute utility measurement for social decision making. *IEEE Transactions on System, Man and Cybernetics*, SMC 7-5.

Emmanji, D., Furuta, K. & Kondo, S. (1992). A deterministic model of plant operators using associative network and loose truth maintenance system. *ANP'92*. Tokyo.

Endsley, M. R. (1995). Toward a theory of situation awareness in dynamic systems. *Human Factors*, 37/1, pp. 32-64.

Eurocontrol-Agenda (2005). *The Eurocontrol Safety Agenda 2020*. Brussels (Belgium): Eurocontrol.

Eurocontrol-Levelbust (2004). *Level Bust Tool Kit*. Brussels (Belgium): Eurocontrol.

Eurocontrol-Planning (2003). *European Model for Strategic ATM Investment Analysis*. SD/SNP/SPF_ECO/DOC/041. Brussels (Belgium): Eurocontrol.

Eurocontrol-Shape (2005b). The *Impact of Automation on Future Controller Skill Requirements-A Framework and Toolkit for SHAPE*. Brussels (Belgium): Eurocontrol.

Eurocontrol-Tokai (2003). *Toolkit for ATM Occurrence Investigation*. Brussels (Belgium): Eurocontrol.

Evans, J. (1989). *Bias in Human Reasoning: Causes and Consequences*. Hillsdale (NJ): Erlbaum.

EVEO (2005). *Ereigniserfassung mit dem Programm EVEO*. (see www.linsenmaier.com).

Farmer, E. & Jones, D. (2001). *Applying the Cognitive Streaming Theory to Air Traffic Management. A Preliminary Study, 1: Background and Analysis*. Farnborough (UK): QinetiQ, Center of Human Sciences.

Fastenmeier, W. (1995). *Autofahrer und Verkehrssituationen: Neue Wege zur Bewertung von Sicherheit und Zuverlässigkeit moderner Strassenverkehrssysteme*. Band 33. Köln: TÜV-Rheinland.

Festinger. L. (1957). *A Theory of Cognitive Dissonance*. Stanford: Stanford University Press.

Fitts, P. M. (1954). The information capacity of the human system in controlling the amplitude of movement. *Journal of Experimental Psychology*, 47, pp. 381-391.

Fitts, P. M. (1964). Perceptual motor skills learning. In: Melton, A. W. (Ed), *Categories of Human Learning*. New York: Academic Press, pp. 243-285.

Fraser, J., Smith, P. & Smith, J. (1992). A catalog of errors. *International Journal of Man-Machine Studies*, 37(3), pp. 265-307.

Freud, S. (1936). *Das Ich und die Abwehrmechanismen*. München: Kindler.

Fu, L. M. (1993). Knowledge based connectionism for revising domain theories. *IEEE Transactions on System, Man, and Cybernetics*, 23-1.

Fujita, Y. & Hollnagel, E. (2004). Failures without errors: Quantification of context in HRA. *Reliability Engineering and System Safety*, 83/2 [special issue on human reliability data issues and errors of commission].

Fukuda, R. (2004). *Ergonomische Gestaltung der Webauftritte*. München: Utz.

Fukuda, R. & Voggenberger, T. (2004). *Group Interaction in High Risk Environments Communication in Nuclear Power Plants (Phase 2)*. Cologne (Germany): GItb.

Fukuda, R., Voggenberger, T., Sträter, O. & Bubb, H. (2003). Measuring the quality of group interaction in nuclear power plant environments by using eye movement behaviour. *Ergonomics in the Digital Age: Proceedings of the 15th Triennial Congress of the International Ergonomics Association and the 7th Joint Conference of Ergonomics Society of Korea.* Japan Ergonomics Society.

Furuta, K. (2000). Human factors in JCO criticality accident. In: Kondo, S. & Furuta, K. (Eds), *PSAM 5. Probabilistic Safety and Management.* Tokyo: Universal Academy Press.

Gentner, D. & Stevens, A. (Eds) (1983). *Mental models.* Hillsdale (NJ): Erlbaum.

Gerdes, V. (1997). *Identification and Analysis of Cognitive Errors: Application to Control Room Operators.* Wageningen (NL): Ponsen & Looijen.

Gertman, D. I., Blackman, H. S., Haney, L. N., Deidler, K. S. & Hahn, H. A. (1992). INTENT: A method for estimating human error probabilities for decision-based errors. *Reliability Engineering and System Safety*, 35, pp. 127-136.

Gibson, J. (1973). *Die Wahrnehmung der visuellen Welt.* Weinheim: Beltz.

GIHRE (2004). *Group Interaction in High Risk Environments.* Aldershot (UK): Ashgate.

Gillard, P. (1979). Interventionsmöglichkeiten im Projektstadium zur Erreichung eines guten akustischen Komforts. *Unikeller Konferenzbericht.*

Grote, G. (1997). *Autonomie und Kontrolle: Zur Gestaltung automatisierter und risikoreicher Systeme.* Zürich (CH): vdf Hochschulverlag.

GRS (1998a). *Insights from the Analysis of the Impact of Organizational Behaviour on Safety Performance in a German NPP.* Cologne (Germany): GRS [see also www.grs.de].

GRS (1998b). *Untersuchungen der Sicherheitsreserven von Kernkraftwerken bei auslegungs-überschreitenden Ereignissen (Accident Management).* SR 2223. Cologne (Germany): GRS.

GRS (1999). *SWR-Sicherheitsanalyse Phase II: Untersuchungen von Ereignissen außerhalb der Leistungsbetriebs.* Cologne (Germany): GRS.

Guski, R. (1989). Wahrnehmung: Eine Einführung in die Psychologie der menschlichen Informationsaufnahme. In: Selg, H. & Ulich, D. (Eds), *Grundriß der Psychologie*, 7. Stuttgart: Kohlhammer.

Hannaman, G. W. & Spurgin, A. J. (1984). *Human Cognitive Reliability Model for PRA Analysis.* NUS-4531. San Diego: NUS-Corporation.

Häusler, R. (2005). *Teamprozesse effektiver Cockpit-Crews: Adaptation an spezifische Aufgaben- und Situationsanforderungen.* [Dissertation, Institut für Psychologie, Universität Bern].

Hebb, D. O. (1949). *The Organization of Behaviour.* New York: Wiley,

Helmreich, R. L. & Merritt, A. C. (1998). *Culture at Work in Aviation and Medicine: National, Organizational, and Professional Influences.* Aldershot (UK): Ashgate.

Hick, W. (1952). On the rate of gain of information. *Quarterly Journal of Experimental Psychology*, 4, pp. 11-26.

Hockey, R. (1984). *Stress and Fatigue.* New York: Wiley.

Hofstätter, P. (1973). *Sozialpsychologie.* Berlin: de Gryter.

Hohlfeld, A., Sangals, J. & Sommer W. (2004). Effects of additional tasks on language perception: An ERP investigation, *Journal of Experimental Psychology: Learning, Memory, Cognition.* <http://www.agnld.uni-potsdam.de/~romano/Promotionskolleg/dimigen2.pdf> (01.02.2005).

Hollnagel, E. (1992). The reliability of man-machine interaction. *Reliability Engineering and System Safety*, 38, p. 81 ff.

Hollnagel, E. (1993). *Human Reliability Analysis: Context and Control.* London: Academic Press.

Hollnagel, E. (.1998). *Cognitive Reliability and Error Analysis Method: CREAM.* New York: Elsevier.

Hollnagel, E. (1999). Looking for errors of omission and commission or the hunting of the snark revisited. *Reliability Engineering and System Safety.*

Hollnagel, E. (2004). *Barriers and Accident Prevention.* Aldershot (UK): Ashgate.

Hollnagel, E. & Amalberti, R. (2001). The Emperor's New Clothes, or whatever happened to 'human error'. *4th International Workshop on Human Error, Safety and System Development*, Linköping University, June 11-12, 2001, Linköping.

Holy (2004). Some insights from recent applications of HRA methods in PSA effort and plant operation feedback in Czech Republic. *Reliability Engineering and System Safety*, 83/2 [special issue on human reliability data issues and errors of commission].

Husserl, E. (1976). *Ideen zur reinen Phänomenologie und phänomenologischen Philosophie.* Den Haag: Hussrliana.

IAEA (1990). *Human Error Classification and Data Collection.* TECDOC 538. Vienna: IAEA.

IAEA (2001a). *IRS Study on Incidents Caused by Loss of Corporate Knowledge And Memory. Phase II: In depth analysis of selected events.* IAEA-J4-CS-04/01. Vienna: IAEA.

IAEA (2001b). *Guidelines for Describing of Human Factors in the IRS.* IAEA-J4-CS-10. Vienna: IAEA.

IAEA (2003). *IAEA Expert Review Mission Completes Assessment of Fuel Cleaning Incident at Paks Nuclear Power Plant.* Vienna: IAEA.

ICAO (1989). ADREP Summary. Aviation Week & Space Technology, *Flight Int.* 17-12.01.1990 p. 44 [see also <http://aviation-safety.net/database/1989/890219-0.htm> (01.02.2005)].

Isaac, A. Shorrock, S. Kennedy, R., Kirwan, B., Andersen, H. & Bove, T. (2003). *The Human Error in ATM Technique (HERA JANUS).* HRS/HSP-002-REP-03. Brussels (Belgium): Eurocontrol.

Isaac, A., Sträter, O. & Van Damme, D. (2004). *A Method for Predicting Human Error in ATM (HERA Predict).* HRS/HSP-002-REP-07. Brussels (Belgium): Eurocontrol.

ISO-10075 (2002). *Ergonomic Principles Related to Mental Workload.* Geneve (CH): International Organization for Standardization.

ISO-9241 (1993). *Ergonomic Requirements for Office Work with Visual Display Terminals (VDTs).* Geneve (CH): International Organization for Standardization.

Jeannot, E., Kelly, C. & Thompson, D. (2003). *The Development of Situation Awareness Measures in ATM Systems*. HRS/HSP-005-REP-01. Brussels (Belgium): Eurocontrol.

Johnson-Laird, P. (1983). *Mental Models*. Cambridge: Harvard University Press.

Jones, D. & Farmer, E. (2001). *Applying the Cognitive Streaming Theory to Air Traffic Management: A Preliminary Study*. Brussels (Belgium): Eurocontrol.

Jones, D., Macken, W. & Nicholls, A. (2003). *The Phonological Store of Working Memory Reconsidered*. MS 02-108-MS 2003-0039, Cardiff: School of Psychology of the Cardiff University.

Kahneman, D. & A. Tversky (1979). Prospect Theory: An Analysis of Decision under Risk. *Ecometrica*, 47, pp. 263-291.

Kanse, L. & van der Schaaf, T. (2000). Recovery from failures-understanding the positive role of human operators during incidents. In: Waard, D., Weikert, C. Hoonhout, J. & Ramaekers, J. (Eds), *Human System Interaction: Education, Research and Application in the 21st Century*. Maastricht (NL): Shaker Publishing.

Kant, I. (1966). *Kritik der reinen Vernunft*. Stuttgart: Reclam.

Katzenberger, L. F. (1967). *Auffassung und Gedächtnis*. München: Ernst Reinhardt Verlag.

Keller, A. (1990). *Allgemeine Erkenntnistheorie*. Stuttgart: Kohlhammer.

Keller, R. M. (1987). *The Role of Explicit Contextual Knowledge in Learning Concepts to Improve Performance*. [PhD Thesis, Rutgers University, Computer Science Department, New Jersey] (Technical Report, ML-TR-7).

Kelly, C. (2003). *Guidelines for Trust in Future ATM Systems: Principles*. HRS/HSP-005-GUI-03. Brussels (Belgium): Eurocontrol.

Kelly, C., Boardman, M., Goillau, P. & Jeannot, E. (2003). *Guidelines for Trust in Future ATM Systems: A Literature Review*. HRS/HSP-005-GUI-01. Brussels (Belgium): Eurocontrol.

Kempke, C. (1988). Der neuere Konnektionismus: Ein Überblick. *Informatik Spektrum*, 11, pp. 143-162.

Kerninghan, B. & Ritchie, D. (1988). *The C Programming Language*. Englewood Cliffs (NJ): Prentice Hall.

Kieras, D. E. & Polson, R. G. (1985). An approach to the formal analysis of user complexity. *International Journal of Man-Machine Studies*, 4, pp. 365-394.

King, P. M. & Kitchener K. S. (1994). *Developing Reflective Judgment: Understanding and Promoting Intellectual Growth and Critical Thinking in Adolescents and Adults*. San Francisco: Jossey-Bass.

Kirwan, B. (2002). *Human Factors Impacts in Air Traffic Management: Some Examples, Human Factors for Project Managers*, Brussels (Belgium) Eurocontrol [Training Course Material].

Kirwan, B. & Straeter, O. (2002). The Grass is Always Greener: Contrasting Safety Management in Nuclear Power and Air Traffic Management. *Human Factors and Safety in Aviation*, Lund University, September 26–27, 2002, Lund (Sweden).

Kirwan, B., Gibson, H. Kennedy, R., Edmunds, J., Cooksley, G. & Umbers, I. (2004). Nuclear Action Reliability Assessment (NARA). A Data-Based HRA Tool. In: Spitzer, C., Schmocker, U. & Dang, V. (Eds). *Probabilistic Safety Assessment and Management. Proceedings of the PSAM 7 2004.* Berlin: Springer.

Kohda, T., Nojiri, Y. & Inoue, K. (2000). Root cause analysis of JCO accident based on decision-making Model. In: Kondo, S. & Furuta, K. (Eds), *PSAM 5. Probabilistic Safety and Management.* Tokyo: Universal Academy Press.

Köhler, T. (2001). *Anwendung und Weiterentwicklung von CREAM am Beispiel der Bedienung und Überwachung einer verfahrenstechnischen Anlage.* [Diploma Thesis, Institut für Psychologie und Arbeitswissenschaft, Technische Universität zu Berlin].

Kohonen, T. (1988). *Self- Organization and Associative Memory.* Berlin: Springer.

Kontogiannis, T. & Lucas, D. (1990). *Operator Performance under High Stress: An Evaluation of Cognitive Modes, Case Studies and Countermeasures.* Report R 90/03. Tokyo: Nuclear Power Engineering Test Centre [available from Human Reliability Associates, Dalton].

Kosslyn, S. M. (1973). Scanning visual images: Some structural implications. *Perception and Psychophysics,* 14, pp. 90-94.

Krüger, H. (1998). *Lokale Feldpotentiale im Elektrokortikogramm und Elektroenzehpalogramm des Menschen: Nachweis, Beschreibungskriterien, Anwendung.* [Dissertation, Medizinische Fakultät Charité, Humboldt-Universität, Berlin].

Kuhn, T. (1976). *Die Struktur wissenschaftlicher Revolutionen.* Frankfurt am Main: Suhrkamp Verlag.

Kuhn, W., Fricke, B., Schäfer, K. & Schwarz, G. (1976). *Quantenphysik.* Braunschweig: Westermann.

LeBot, P. (2004). Human reliability data, human error and accident models: Illustration through the Three Mile Island accident analysis. *Reliability Engineering and System Safety,* 83/2 [special issue on human reliability data issues and errors of commission].

LeBot, P., Cara, F. & Bieder, C. (1999). MERMOS: A second generation HRA Method. What it does and doesn't do. In: Modarres, M. (Ed), *PSA'99. Risk-Informed Performance-Based Regulation.* LaGrange Park (Illinois): American Nuclear Society.

Leeuwenberg, E. L. J. (1968). *Structural Information of Visual Patterns.* The Hague: Mouton & Co, pp. 5-36.

Leveson, N. (2002). *A New Approach to System Safety Engineering.* Boston: Massachusetts Institute of Technology.

Linsenmaier (2003). Achieving consistent and detailed incident analyses for a common interdisciplinary human reliability data-base: A study in aviation. *IEA 2003, Seoul (Korea)*

Linsenmaier, B. (2005). *Abbildung von Ereignissen mit menschlichen Handlungsfehlern auf ein Mensch-Maschine Modell.* [Dissertation, Fakultät für Maschinenwesen, Technische Universität München].

Linsenmaier, B. & Sträter, O. (2000). Recording and evaluation of human factor events with a view to system awareness and ergonomic weak points within the system at the example of commercial aeronautics. In: HFES (Ed), *Human Factors and Ergonomics Society*. St. Lois: Mira Digital Publishing.

Lorenz, K. (1978). *Vergleichende Verhaltensforschung: Grundlagen der Ethologie*. Berlin: Springer.

Low, I. (2004). *A Tool for the Assessment of the Impact of Change in Automated ATM Systems on Mental Workload*. HRS/HSP-005-REP-03. Brussels (Belgium): Eurocontrol.

Malsburg, C. von (1986). Am I thinking in assemblies. In: Palm, G. & Aertsen, A. (1982), *Brain Theory*. Heidelberg: Springer.

Mandl, H. & Spada, H. (1988). *Wissenspsychologie*. München: Psychologie Verlagsunion.

McClelland, J. & Rummelhart, D. (1981). An interactive activation model of context effects in letter perception. *Psychological Review*, 88/4, pp. 375-405.

McCormick, E. & Sanders, M. (1983). *Human Factors in Engineering and Design*. Auckland: McGraw-Hill.

McRuer, D. T. (1967). A review of quasi-linear pilot models. *IEEE Transactions on Human Factors in Electronics*. HFE-8, p. 231.

Mechsner, F. (1998). Die Suche nach dem Ich. *Geo*, 2/1998, p. 62 ff.

Mehl, R. (1995). *Zuverlässigkeit von komplexen Systemen*. Arbeitspapier der Arbeitsgruppe Software-Zuverlässigkeit, Mai 1995, VDI, Düsseldorf.

Meister, D. & Hogg, D. N. (1995). *Development of a Task Descriptive Model of the Operator's Fault Diagnosis: A Framework for Interpreting Empirical Data within the Human Error Project*. HWR 379. Halden (Norway): OECD Halden Reactor Project.

Messing, J. (1999). *Allgemeine Theorie des menschlichen Bewusstseins*. Berlin: Weidler Buchverlag.

Michel, T. (1989). *Über die Philosophischen Grundlagen der Wissenschaft vom Menschen*. [Dissertation, Philosophische Fakultät, Universität Saarbrücken].

Miller, G. A. (1956). The magical number seven plus or minus two: Some limits on our capacity for processing information. *Psychological Review*, 63, p. 81.

Milner, R. & Michalski, A. (2003). Cortical responsiveness is reduced during P300 potential: Does the level of initial activity affect this inhibition. *Acta Neurobiol. Exp. 2003*, 63, pp. 351-360.

Mitchell, J. & Rahmann, M. (2004). Critical incident stress management: Theory and background. *26th conference of the European Aviation Psychology (EAAP)*, EAAP, 2004, Lisbon.

Moieni, P., Spurgin, A. J. & Singh, A. (1994). Advances in human reliability analysis methodology. Part 1: Frameworks, models and data. *Reliability Engineering and System Safety*, 44, p. 27.

Montmollin, M. (1992). Activity-oriented ergonomics: Models and methods. *Rencontre Européenne organisée avec le soutien du Département Etude de Sureté et de Fiabilité de la Direction des Etudes et Recherches de L'EDF et l'Accociation Naturalia & Biologia*, EDF, 1992, Paris.

Moray, N. (1987). Intelligent aids, mental models and the theory of machines. *International Journal of Man-Machine Studies*, 27, pp. 619-629.

Moray, N. (1990). A lattice theory approach to the structure of mental models. In: Broadbent, D., Reason, J. & Baddeley, A. (Eds), *Human Factors in Hazardous Situations*. Oxford: Clarendon Press, pp. 129-135.

Mosleh, A. & Chang, Y. (2004). Model-based human reliability analysis: Prospects and requirements. *Reliability Engineering and System Safety*, 83/2 [special issue on human reliability data issues and errors of commission].

Mosneron-Dupin, F., Reer, B., Heslinga, G., Sträter, O., Gerdes, V., Saliou, G. & Ullwer, W. (1997). Human-centered modelling in human reliability analysis: Some trends based on case studies. *Reliability Engineering and System Safety*, 58/3, pp. 249-274.

Müller-Ecker, D., Mayer, G. & Sträter, O. (1998). Probabilistic safety assessment for non-full-power states of NPPs in Germany. In: Kondo, S. & Furuta, K. (Eds), *PSAM 5. Probabilistic Safety and Management*. Tokyo: Universal Academy Press.

Münte, T. (2002). Irren ist menschlich: Befunde zur Fehlerverarbeitung im Gehirn. *Das gläserne Gehirn: Moderne Bildgebung. Der Schlüssel zum menschlichen Bewusstsein,* Wissenschaftszentrum Nordrhein-Westfalen, 2002/03, Düsseldorf.

Nauck, D., Klawonn, F. & Kruse, R. (1994). *Neuronale Netze und Fuzzy-Systeme: Grundlagen des Konnektionismus, neuronaler Fuzzy-Systeme und der Kopplung mit wissensbasierten Methoden*. Braunschweig: Vieweg.

Neisser, U. (1976). *Cognition and Reality*. San Francisco: W. H. Freeman.

Neumann, O. (1992). Zum gegenwärtigen theoretischen Umbruch in der Kognitionspsychologie. *Merkur-Zeitschrift für europäisches Denken*, 1992/1, p. 48.

Newell, A. (1990). *Unified Theories of Cognition*. Cambridge: Harvard University Press.

Norman, D. A. (1981). Categorizing of action slips. *Psychological Review*, 88, pp. 1-14.

Norman, D. A. (1986). New views on information processing: Implications for intelligent decision support systems. In: Hollnagel, E., Mancini, G. & Woods, D. D. (Eds), *Intelligent Decision Support in Process Environments*. Berlin: Springer.

NUREG-1624 (2000). *Technical Basis and Implementation Guidelines for A Technique for Human Event Analysis (ATHEANA)*. Rev 1, NUREG-1624. Washington DC: NRC.

OECD (2000). *Errors of Commission in Probabilistic Safety Assessment*. NEA/CSNI/R(2000), 17. Paris: OECD/NEA.

OECD (2001). *Building the New HRA*. Workshop of the OECD/NEA & NRC. Washington DC: NRC.

OECD (2002). *Strengthening the Link between HRA and Data*. Workshop of the OECD/NEA & GRS, GRS, Munich (Germany).

Parasuraman, R., Molloy, R. & Singh, I. L. (1993). Performance consequences of automation induced complacency. *International Journal of Aviation Psychology*, 3, pp.1-23.

Park, J., Jung, W., Ha, J. & Shin, Y. (2004). Analysis of operators' performance under emergencies using a training simulator of the nuclear power plant. *Reliability Engineering and System Safety*, 83/2 [special issue on human reliability data issues and errors of commission].

Pauli, R. & Arnold, W. (1957). *Psychologisches Praktikum*. Stuttgart (Germany): Fischer.

Payne, J. (1980). Information processing theory: Some concepts and methods applied to decision research. In: Wallsten (Ed), Cognitive *Processes in Choice and Decision Behavior*. Hillsdale (NJ): Erlbaum.

Pearl, J. (1988). *Probabilistic Reasoning in Intelligent Systems-Networks of Plausible Inference*. San Mateo (California): Morgan Kaufmann Publishers.

Perrow, C. (1984). *Normal Accidents: Living with High Risk Technologies*. NY (US): Basic Books 1984.

Piaget, J. (1947). *Psychologie der Intelligenz*. Zürich (CH): Rascher.

Pope, D., Houghton, R., Jones, D. & Parmentier, F. (2003). *Cognitive Streaming Project. Report on Work Package 2: Range of Tasks Affected by Irrelevant Sound*. CARE-IA-CS-CFU-WP2-D2-02-B, Brussels (Belgium): Eurocontrol.

Popper, K. R. (1997). *Auf der Suche nach einer besseren Welt*. 9. Auflage, München: Piper.

Posner, M. I. & Synder, C. R. R. (1975). Attention and cognitive control. In: Solso, R. (Ed), *Information Processing and Cognition: The Loyola Symposium*. Hillsdale: Erlbaum.

Potash, L., Stewart, M., Dietz, P. Lewis, C. & Dougherty, E. (1981). Experience in integrating the operator contributions in the PRA in actual operating plants. *ANS/ENS Topical Meeting on Probabilistic Risk Assessment*, American Nuclear Society, September 1981, La Grange Park (US), Vol. II, pp. 1054-1063.

Puppe, F. (1988). *Einführung in Expertensysteme*. Berlin: Springer.

Pyy, P. (2001). The effect of organizational factors on risk. In: Kafka, P. (Ed), *PSA RID: Probabilistic Safety Assessment in Risk Informed Decision Making*. European Commission, 4.-9.3.2001, GRS, Munich (Germany) [EURO-Course].

Rasch, G. (1980). *Probabilistic Models for Some Intelligence and Attainment Tests*. Chicago: University of Chicago Press.

Rasmussen, J. (1982). The role of cognitive models of operators in the operation and licensing of nuclear power plants. In: Sheridan, T. S., Jenkins, J. P., Kisner, R. A. & Abbott, L. S. (Eds), *Proceedings of Workshop on Cognitive Modelling of Nuclear Power Plant Control Room Operators*. NUREG/CR-3114, NRC, Washington DC, p. 13 ff.

Rasmussen, J. (1986). *Information Processing and Human-Machine Interaction*. New York: North-Holland.

Rassl, R. (2004). *Ablenkungswirkung tertiärer Aufgaben im PKW: Systemergonomische Analyse und Prognose.* [Dissertation, Technische Universität München].

Reason, J. (1990). *Human Error.* Cambridge: Cambridge University Press.

Reason, J. (1997). *Managing the Risk of Organizational Accidents.* Aldershot (UK): Ashgate.

Reer, B. (1993). *Entscheidungsfehler des Betriebspersonals von Kernkraftwerken als Objekt probabilistischer Risikoanalysen.* [Dissertation, KFA Jülich].

Reer, B. (1995). *Analyse menschlicher Zuverlässigkeit in technischen Systemen.* KFA-ISR-IB 1/95. Jülich (Germany): KFA-Jülich.

Reer, B., Bongartz, R. & Ullwer, W. (1995). *Zuverlässigkeitsanalyse von Personalhandlungen bei Störungen des Leistungsbetriebes eines Siedewasserreaktor: Insbesondere von Notfallmaßnahmen.* KfA-IST-IB-7/95. Jülich (Germany): KFA-Jülich.

Reer, B., Sträter, O., Dang, V. & Hirschberg, S. (1999). *A Comparative Evaluation of Emerging Methods for Errors of Commission Based on Applications to the Davis-Besse Event 1985.* PSI, Schweiz, Nr. 99-11.

Reichart, G. (2000). *Menschliche Zuverlässigkeit beim Führen von Kraftfahrzeugen-Möglichkeiten der Analyse und Bewertung.* [Dissertation, Fakultät für Maschinenwesen, Technische Universität München].

Rigby, L. (1970). The nature of human error. *Annual technical Conference Transactions of the ASQC,* ASQC, Milwaukee.

Roelen, A. L. C., Wever, R. & Verbeek, M. (2003). *Improving Aviation Safety by Better Understanding and Handling of Interfaces: A Pilot Study.* NLR-CR-2003, DGL/2.03.82.129. Amsterdam (NL): NLR.

Roessingh, J. & Zon, R. (2004). *A Measure to Assess the Impact of Automation on Teamwork.* HRS/HSP-005-REP-07. Brussels (Belgium): Eurocontrol.

Roth, E. (1967). *Einstellung als Determination individuellen Verhaltens.* Göttingen (Germany): Hogrefe.

Rothaug, J. (2003). *Age, Experience and Automation in European Air Traffic Control.* HRS/HSP-005-REP-02. Brussels (Belgium): Eurocontrol.

Rouse, W. & Rouse, S. (1983). Analysis and classification of human error. *IEEE Transactions on Systems. Man and Cybernetics,* 4, pp. 539-549.

Rummelhart, D. & McClelland, J. (Eds) (1986). *Parallel Distributed Processing.* Cambridge (MA): MIT Press.

Rummelhart, D. E. & Norman, D. A. (1982). Simulating a skilled typist: A study of skilled cognitive-motor performance. *Cognitive Science,* 6, pp. 1-36.

Rummelhart, D., Lindsay, P., Norman, D. (1972). A process model for long-term memory. In: Tulving E., Donaldsen W. (Eds), *Organization of Memory.* New York: Academic Press 1972.

Sanders, A. F. (1983). Towards a model of stress and human performance. *Acta Psychologica,* 53, pp. 61-97.

Sanders, A. F. (1975). Some remarks on short-term memory. In: Rabbitt, P. M. A. & Dornic, S. (Eds), *Attention and Performance.*

Sangals, J., Sommer, W. & Leuthold, H. (2002). Influences of presentation mode and time pressure on the utilisation of advance information in response preparation. *Acta Psychologica*, 109, pp. 1-24.

Sarbin, T. R. & Bauley, D. E. (1966). The immediacy postulate in the light of modern cognitive psychology. In: Hammond, K. R. (Ed), (1966), *The Psychology of Egon Brunswick.* New York: Holt, Rinehart & Winston, p. 159 ff.

Schäfer, D, Meckiff, C., Magil, A., Picard, B. & Aligne, F. (2001). Air traffic complexity as a key concept for multi-sector planning. *DASC 2001.*

Schandry, R. (1981). *Psychophysiologie. Körperliche Indikatoren menschlichen Verhaltens.* München: Urban & Schwarzenberg.

Schemmerer, U. (1993). *Mentales Fahrermodell für das Spur- und Abstandhalten.* [Diploma Thesis, Philosophisch-Pädagogische Fakultät, Universität Eichstätt].

Schmidt, R. A. (1975). A schema theory of discrete motor skill learning. *Psychological Review*, 82, pp. 255-260.

Schmidt, R. A. (1988). *Motor Control and Learning: A Behavioural Emphasis.* Champaign (Illinois): Human Kinetics Publishers.

Schmidt, R.F. & Thews, G. (1987). *Physiologie des Menschen.* 23. Auflage. Heidelberg: Springer.

Schmidtke, H. (1993). *Ergonomie.* München: Hanser.

Schneider, W. & Shiffrin, R. W. (1977). Controlled and automatic human information processing: Decision, search, and attention. *Psychological Review*, 84, pp. 1-66.

Schnotz, W. (1988). Textverstehen als Aufbau mentaler Modelle. In: H. Mandl & H. Spada (Hg.), *Wissenspsychologie.* München (Germany): Psychologie Verlags Union, pp. 299-330.

Schuler, H. (Ed) (1995). *Lehrbuch Organizationspsychologie.* Bern (CH): Hans Huber.

Schweigert, M. (1999). *Fahrerverhalten beim Führen eines Kraftfahrzeuges unter gleichzeitiger Bearbeitung von Zusatzaufgaben.* Lehrstuhl für Ergonomie der TU München [Abschlussbericht an die Adam Opel AG, BMW AG, DaimlerChrysler AG, Robert Bosch GmbH, BmB+F-Projekt „MoTiV"].

Schweigert, M. (2003). *Fahrerblickverhalten und Nebenaufgaben.* [Dissertation, Fakultät für Maschinenwesen, Technische Universität München].

Schweigert, M., Fukuda, R. & Bubb, H. (2001). Blickerfassung mit JANUS II. In: GFA (Ed), *Arbeitsgestaltung, Flexibilisierung, Kompetenzentwicklung.* Dortmund (Germany): GfA Press.

Selfridge, O. G. (1955). Pattern recognition in modern computers. *Proceedings of the Western Joint Computer Conference.*

Semmer, N. (1994). Der menschliche Faktor in der Arbeitssicherheit: Mechanismen, Verhütung und Korrektur von menschlichen Fehlhandlungen. *Ursachenanalyse von Störfällen in Kernkraftwerken*, SVA, 3.3.1994, Brugg-Windisch (CH) [Ausbildungsseminar].

Shastri, L. (1988). A connectionist approach to knowledge representation and limited inference. *Cognitive Science*, 12, pp. 331-392.

Shastri, L. & Ajjanagadde, V. (1990). *From Simple Associations to Systematic Reasoning: A Connectionist Representation of Rules, Variables and Dynamic Bindings*. MS-CIS-90-05, LINC LAB 162, University of Pennsylvania, Department of Computer and Information Science, School of Engineering and Applied Science, Philadelphia.

Shearer, D., Emmerson, R. & Dustman, R. (2004). *EEG Changes in the Normal Elderly: A Brief Review*. Neuro-psychology Research Laboratories, Veterans Affairs Medical Center, Salt Lake City (Utah).

Shell (2002). *Managing Breaking Rules*. Shell, Den Haag.

Shepherd, C. (2002). *Risk Monitors: A Report on the State of the Art in their Development and Use*. OECD & IAEA, WG RISK. Paris: OECD NEA. [prepared by the UK Nuclear Installations Inspectorate]

Sheridan, T. B. (2002). *Human and Automation: System Design and Research Issues*. New York: Wiley.

Shneiderman, B. (1998). *Designing the Users Interface*. Reading (MA): Addison Wesley.

Shorrock, S. & Sträter, O. (2004). Managing System Disturbances in Air Traffic Management: Elaboration of a Conceptual Framework. *HFESA 2004*, HFESA, Cairns (Australia).

Shorrock, S. T. (1997). *The Development and Evaluation of TRACER: A Technique for the Retrospective Analysis of Cognitive Errors in Air Traffic Control*. [MSc Thesis, The University of Birmingham].

Siemens (2002). *Jahresbericht Arbeitssicherheit 2002*. München (Germany): Siemens.

Simon, H. (1955). A behavioural model of rational choice. *Quarterly Journal of Economics*, 69, pp. 129-138.

Siu, N., (1994). Risk assessment for dynamic systems: An overview. *Reliabililty Engineering and System Safety*, 43, pp. 43-73.

Skyway (2002). *Human Factors in Air Traffic Control*. 6, 24. Brussels (Belgium): Eurocontrol.

Skyway (2004). *The Single European Sky*. 7, 32. Brussels (Belgium): Eurocontrol.

Sommer, W., Leuthold, H., & Matt, J. (1998). The expectancies that govern the P300 amplitude are mostly automatic and unconscious. *Behavioural and Brain Sciences*, 21, pp. 149-150.

Spanner, B. (1993). *Einfluß der Kompatibilität von Stellteilen auf die menschliche Zuverlässigkeit*. VDI Reihe Biotechnik, Nr. 89. Düsseldorf (Germany): VDI Verlag.

Spitzer, C., Schmocker, U. & Dang, V. (2004). Probabilistic safety assessment and management. *Proceedings of the PSAM 7 2004*. Berlin: Springer.

Spurgin, A. (2004). Developments in HRA technology from nuclear to aerospace. In: Spitzer, C., Schmocker, U. & Dang, V. (Eds), *Probabilistic Safety Assessment and Management. Proceedings of the PSAM 7 2004*. Berlin: Springer.

Sternberg, S. (1969). On the discovery of processing stages: Some extensions of Donders' method. *Acta Psychologica*, 35, pp. 276-315.

Bibliography 259

Stoffer, Th. (1989). Perspektiven konnektionistischer Modelle: Das neuronale Netzwerk als Metapher im Rahmen der kognitionspsychologischen Modellbildung. In: Meinecke, C. & Kehrer, L. (Eds) *Bielefelder kognitionspsychologische Beiträge*. Göttingen (Germany): Hogrefe.

Störig, H. J. (1988). *Kleine Weltgeschichte der Philosophie*. 14. Auflage, Stuttgart (Germany): Kohlhammer.

Sträter, O. (1991). *Ein computerunterstütztes Strukturlegeverfahren zur objektorientierten Erfassung von Diagnosestrategien*. [Diploma Thesis, Institut für Psychologie, RWTH Aachen].

Sträter, O. (1994). An expert knowledge oriented approach for the evaluation of the man-machine interface. In: Ruokonen, T. (Ed), *Fault Detection, Supervision and Safety for Technical Processes: SAFEPROCESS '94*. Vol. 2, Helsinki (Finland): Finnish Society of Automation, p. 673 ff.

Sträter, O. (1997). *Beurteilung der menschlichen Zuverlässigkeit auf der Basis von Betriebserfahrung*. GRS-138. Cologne (Germany): GRS.

Sträter, O. (1997b). Investigations on the influence of situational conditions on human reliability in technical systems. In: Seppälä, P., Luopajärvi, T., Nygard, C. & Mattila, M. (Eds), *Proceedings of the 13th Triennial Conference of the International Ergonomic Association*. IEA, June 1997, Tampere (Finland). Vol. 3, p. 76 ff.

Sträter, O. (1998). Problems of cognitive error quantification and approaches for solution. In: Kondo, S. & Furuta, K. (Eds), *PSAM 5. Probabilistic Safety and Management*. Tokyo: Universal Academy Press.

Sträter, O. (2000). *Evaluation of Human Reliability on the Basis of Operational Experience*. GRS-170. Cologne (Germany): GRS [Translation of Sträter, O. (1997) (see also www.grs.de)].

Sträter, O. (2001). The quantification process of human interventions. In: Kafka, P. (Ed), *PSA RID: Probabilistic Safety Assessment in Risk Informed Decision Making*, European Commission, 4.-9.3.2001, GRS, Munich (Germany) [EURO-Course].

Sträter, O. (2001). Modelling and assessment of cognitive aspects of the task of a driver. In: VDI (Ed), *The Driver in the 21st Century*. No. 1613 VDI. Düsseldorf: VDI.

Sträter, O. (2002). *Group Interaction in High Risk Environments: Communication in NPP*. GRS Report A-3020. Cologne (Germany): GRS.

Sträter, O. (2004). Implementation and monitoring of the effectiveness of safety management systems: Approaches and challenges in air traffic control and nuclear energy. *Sicherheitsmanagement in der Kerntechnik* (Symposium), TÜV Süddeutschland, 5.-6. Oktober 2004, München.

Sträter, O. (2004a). Considerations on the elements of quantifying human reliability. *Reliability Engineering and System Safety*, 83/2 [special issue on human reliability data issues and errors of commission].

Sträter, O. (Ed) (2004b). Human reliability data issues and errors of commission. *Reliability Engineering and System Safety*, 83, 2.

Sträter, O. & Zander, R. M. (1998). Approaches to more realistic risk assessment of shutdown states. In: OECD (Ed), *Reliability Data Collection for Living PSA*. NEA/CSNI/R (98), 10. Paris: OECD NEA, p. 236 ff.

Sträter, O. & Bubb, H. (1998). Assessment of human reliability based on evaluation of plant experience: requirements and their implementation. *Reliability Engineering and System Safety*, 63, 2, pp. 199-219.

Sträter, O. & Bubb, H. (2003). Design of systems in settings with remote access to cognitive performance. In: Hollnagel, E. (Ed), *Handbook of Cognitive Task Design*. Hillsdale (NJ): Erlbaum.

Sträter, O. & Kirwan, B. (2002). Differences between human reliability approaches in nuclear- and aviation-safety. *IEEE Seventh Conference on Human Factors and Power Plants: New Century New Trends*, IEEE, September 15-19, 2002, Scottsdale (Arizona).

Sträter, O. & Reer, B. (1999). A comparison of the application of the CAHR method to the evaluation of PWR- and BWR-events and some implications for the methodological development of HRA. In: Modarres, M. (Ed), *PSA'99: Risk-Informed Performance-Based Regulation*. LaGrange Park (Illinois): American Nuclear Society.

Sträter, O., & Van Damme, D. (2004). *Retrospective analysis and prospective integration of human factor into safety-management.* In: Goeters, K.-M. (Ed), *Aviation Psychology: Practice and Research*. Aldershot (UK): Ashgate.

Sträter, O., Schweigert, M. & Fraczek, I. (2001). The impact of errors of commission on human reliability in a car-driving task. In: Zio, E. Demichela, M. & Piccini, N. (Eds), *Esrel 2001: Towards a Safer World*. Turin (Italy): Politecnico Di Torino.

Sträter, O., Isaac, A. & Van Damme, D. (2002). Considerations on the elements of the quantification of human reliability. *Strengthening the Link between HRA and Data*, OECD/NEA & GRS, 2002, Munich (Germany).

Sträter, O., Voller, L., Low, I. & Shorrock, S. (2003). Solutions for human automation partnership in European air traffic management. *IEA 2003*. Seoul (Korea).

Sträter, O., Voller, L. & Low, I. (2004a). Towards an integrated human automation management. *Human Computer Interaction,* HCI Aero, 2004, Toulouse.

Sträter, O., Dang, V., Kaufer, B. & Daniels, A. (2004b). On the way to assess errors of commission. *Reliability Engineering and System Safety*, 83/2 [special issue on human reliability data issues and errors of commission].

Strohschneider, St. (1997). *Strategien und Taktiken beim Problemlösen: Eine kulturvergleichende Untersuchung mit der Computersimulation MORO*. DFG-Projekt ‚Kulturvergleichende Untersuchung der Denk- und Handlungsstile beim Problemlösen' (DO 200/15-1). Lehrstuhl für Psychologie, Bamberg. [Arbeitsbericht]

Struhe, G. (1990). Neokonnektionismus: Eine neue Basis für die Theorie und Modellierung menschlicher Kognition. *Psychologische Rundschau,* 41, p. 129.

Subotic, B. & Ochieng, W. (2005). Recovery from equipment failures in air traffic control. *Ergonomics Society Annual Conference 2005*, UK-ES, 5-7 April, Hatfield (UK).

Swain, A. D. (1989). *Comparative Evaluation of Methods for Human Reliability Analysis.* GRS-71. Cologne (Germany): GRS.

Swain, A. D. (1992). *Quantitative Assessment of Human Errors.* GRS, Cologne (Germany) [Protokoll des Vortrages bei der GRS im Juni 1992].

Swain, A. D. & Guttmann, H. E. (1983). *Handbook of Human Reliability Analysis with Emphasis on Nuclear Power Plant Applications.* Sandia National Laboratories, NUREG/CR-1278. Washington DC: NRC.

Theis, I. (2002). *Das Steer-by-Wire System im Kraftfahrzeug: Analyse der menschlichen Zuverlässigkeit.* München: Shaker.

Theis, I. & Sträter, O. (2001). By-wire systems in automotive industry: Reliability analysis of the driver-vehicle-interface. In: Zio, E. Demichela, M. & Piccini, N. (Eds), *Esrel 2001: Towards a Safer World.* Turin (Italy): Politecnico Di Torino.

Thelwell, P. (1994). What defines complexity. In: Robertson, S. A. (Ed), *Contemporary Ergonomics.* London: Taylor & Francis, pp. 89-94.

Thijs, W. (1887). *Fault Management.* [Diploma Thesis, University of Delft].

Thorndike, E. (1922). *The Psychology of Arithmetic.* New York: Macmillan.

Thurstone, L. L. (1927). A law of comparative judgement. *Psychological Review*, 34, pp. 273-286.

Togerson, W. S. (1962). *Theory and Methods of Scaling.* New York: Wiley.

Trimpop, R. (1994). *The Psychology of Risk Taking Behaviour.* Amsterdam (NL): Elsevier.

Trucco, P. & Leva, M. (2004). *A Probabilistic Cognitive Simulation for HRA Studies: PROCO.* Politecnico di Milano, Department of Management, Economics and Industrial Engineering [to be published in Reliability Engineering and System Safety].

Turing, A. M. (1936). On computable numbers. With an application to the Entscheidungsproblem. *Proceedings of the London Mathematical Society*, 42, pp. 230-265.

Tversky, A. & Kahneman, D. (1974). Judgement under uncertainty: Heuristics and biases. *Science*, 184, pp. 1124-1131.

Van Damme, D. (1998). *Human Factors in the Investigation of Accidents and Incidents.* HUM.ET1.ST13.3000-REP-02, Brussels (Belgium): Eurocontrol.

VDI (1999). *Menschliche Zuverlässigkeit: Ergonomische Forderungen und Methoden der Bewertung.* VDI 4006. Berlin: Beuth-Verlag.

Vester (2002). *Die Kunst vernetzt zu denken.* München: DTV.

Voller, L. & Low, I. (2004). *Impact of Automation on Future Controller Skill Requirements and a Framework for their Prediction.* HRS/HSP-005-REP-04, Brussels (Belgium): Eurocontrol.

Waldrop, M. (1988). Soar: A Unified Theory of Cognition. *Science: New Series*, 241, 4863, pp. 296-298.

Weimer, H. (1931). Fehler oder Irrtum. *Zeitschrift für Pädagogische Psychologie*, 32, pp. 48-53.

Wickens, C. D. (1984). *Engineering Psychology and Human Performance.* Toronto: C. E. Merrill Publishing Company.

Wickens, C. D. (1992). *Engineering Psychology and Human Performance.* New York: Harper Collins Publishers.

Wickens, C. D. & Hollands, J. G. (2000). *Engineering Psychology and Human Performance.* Upper Saddle River (NJ): Prentice Hall.

Wiederhold, G. (1981). *Datenbanken: Analyse, Design, Erfahrungen.* München: Oldenbourg.

Williams, J. C. (1986). HEART: A proposed method for assessing and reducing human error. *9th Advances in Reliability Technology Symposium,* University of Bradford (UK).

Wilpert, B., Fahlbruch, B. & Miller, R. (1998). Sicherheit durch organizationales Lernen. *ATW,* 43/1998, 11.

Woods, D. (2003). *Creating Foresight: How Resilience Engineering Can Transform NASA's Approach to Risky Decision-making.* The Ohio State University, Institute for Ergonomics [Testimony on The Future of NASA for Committee on Commerce, Science and Transportation, John McCain, Chair, October 29, 2003].

Woods, D. D., Roth, E. M., & People, H. E. (1987). *Cognitive Environment Simulation: An Artificial Intelligence System for Human Performance Assessment.* NUREG-CR-4862. Washington DC: NRC.

Wortman, D., Duket, S., Seifert, D. Hann, R. & Chubb, G. (1978). *The SAINT User's Manual.* AMRL-TR-77-62, Wright-Patterson Air Force Base (OH).

Yutaka, F. (2000). JCO criticality accident as post-LOCA: Poor structure induced loss of organizational control accident. In: Kondo, S. & Furuta, K. (Eds), *PSAM 5. Probabilistic Safety and Management.* Tokyo: Universal Academy Press.

Zapf, D., Brodbeck, F. C. & Prümper, J. (1989). Handlungsorientierte Fehlertaxonomie in der Mensch-Computer Interaktion. *Zeitschrift für Arbeits- und Organizationspsychologie,* 33, pp. 178-187.

Author Index

Keyword Index